THE MIXER BOOK

THE MIXER BOOK

Margaret Hudson

Elm Tree Books in association with Thorn Domestic
Appliances (Electrical) Limited

CONTENTS

First published in Great Britain 1972
by Elm Tree Books/Hamish Hamilton Ltd
Garden House 57–59 Long Acre London WC2E 9JZ

Second Impression March 1976
Third Impression October 1976
Fourth Impression February 1977
Fifth Impression November 1978
Sixth Impression April 1980

Copyright © 1972 Elm Tree Books/
Hamish Hamilton Limited

ISBN 0 241 02269 X

Book layouts and decorations by
Norma Crockford

Photographs by Paul Radkai

Printed and bound in Great Britain by
Redwood Burn Limited
Trowbridge & Esher

INTRODUCTION

THE MIXER BOOK is to help you, as a food mixer owner, get the best possible use from your machine. In order to achieve this, you must in the first instance ensure that the mixer is always ready for use. Hand mixers are generally supplied with a wall bracket and so are ever ready for use, but table mixers tend to be kept in cupboards where they often remain, forgotten or neglected. It is essential therefore to see that a permanent spot with adjacent power point is found for your mixer; only then will you use it as a matter of course.

THE RECIPES

As the food mixer enables you to carry out food preparation quickly and efficiently, all equipment should be assembled before mixing commences. For this reason, we indicate at the beginning of each recipe what type of machine is involved, and which attachments are required for the preparation. Many of the recipes can be carried out using either table or hand mixer, but in certain cases— bread-making, for instance—only table models are suitable. It must be made clear that a hand mixer with bowl and stand unit does not constitute a table mixer. The bowl and stand are simply designed to relieve the weight on your hand. The

ingredients for each recipe are listed in the order in which they are to be used. This, too, will save time in the preparation of the dish.

BAKING

The variety of recipes will help you to introduce many new and exciting dishes into your everyday catering and, at the same time, provide others for use when that extra-special something is required. There are dishes here which perhaps previously you have only attempted on high days and holidays. You will soon find that with the mixer they are so simply made that they can become part of your regular baking day. Imagine being able to make a whole batch of light, melt-in-the-mouth meringues without even a hint of an aching arm. Your family will soon be demanding them regularly.

MONEY-SAVER

As well as the saving of time and effort, you will be saving money, too. Apart from the fact that it cuts out the need to buy cakes, pastries, bread etc., great savings are made by preparing soups, drinks, cream and ice-cream, to name but a few. You will be surprised to find that a whole jug of

lemonade can be made for a few pence, and the mother with a young family can forget that tinned baby foods exist, as she prepares perfectly smooth purées in seconds using the liquidiser.

The cake, pastry and bread recipes are so simple that it is just as easy to prepare two or three as one, and so if you have a home freezer, make it a rule of the house always to bake sufficient for both eating and freezing.

MIXER ATTACHMENTS

The attachments which come as optional extras with many machines can play a vital role in helping you towards getting the most from your mixer. These attachments should be chosen with great care so as to ensure that the ones purchased really do prove to be of great assistance. The following notes will help to indicate the functions carried out by the various attachments available.

Liquidiser (or Blender): This is an invaluable attachment which chops, crumbs, purées and blends foods in seconds. Its ability to cope with soups, drinks, pâtés, baby foods etc. has to be seen to be believed—a 'must' for any kitchen.

Juice Separator: Takes juices from fruits and vegetables other than the citrus type. You only have to wash and remove any bad parts from the food—the Separator does the rest. It is excellent for preparing clear soups, vitamin-packed fruit and vegetable drinks, and a boon to any amateur wine-maker.

Juice Extractor: To be used for the extraction of juice from citrus fruits. It is ideal for that early morning drink of lemon or grapefruit juice. Leave the fruit in the refrigerator overnight to ensure that the extracted juice is cold and still vitamin-packed.

Coffee Grinder: Enables one to grind the beans just before using, so there is no question of their not being fresh.

Cream Maker: Prepares both cream and ice-cream at a fraction of the purchase price by combining milk and unsalted butter.

Slicer and Shredder: Not only useful for the preparation of crisp unhandled salads, but for grating nuts, cheese, chocolate etc. to add to made up dishes.

Can Opener: Most useful if left permanently on the machine.

Mincer: Copes admirably with both raw and cooked meats. It's a real joy to be able to dispense with the handle and let the motor do the hard work.

Bean Slicer—Pea Sheller: Beans and peas can be processed in minutes—this helps to ensure that they come to the table in prime condition. Home freezer owners would find this useful for dealing with that glut of vegetables that need freezing without delay.

Potato Peeler: The prospect of no more potato peeling should endear this attachment to any housewife.

Colander and Sieve: Ideal for the preparation of all types of fruit and vegetable purées and baby foods.

Although the recipes contained in this book are for use with a food mixer, it does not follow that the introduction of a machine into your home prevents you from using it on your own well-tried favourites. Do not be afraid to experiment—the results will be as good as, if not better, than they have ever been before, and you will not have had to work so hard. Should you not be too sure of the speed setting, then consult the instruction book that came with your machine and follow that; experience will help.

Cooking should always be fun; so go ahead—your mixer is ready, the recipes, which we at Kenwood have tested fully, await you. Happy Mixing!

Margaret Hudson

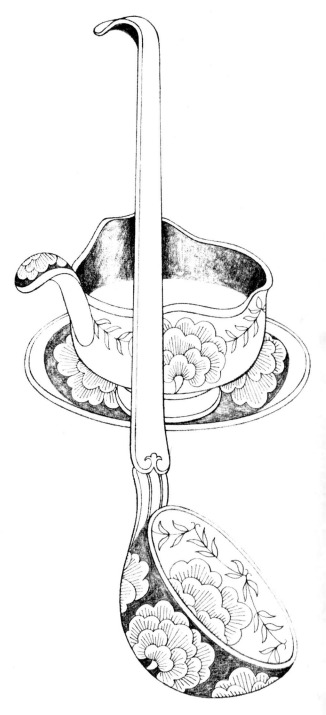

The starter sets the tone of any meal, so it must be just right, yet at the same time not demand so much attention that the main course suffers. Tomato Cocktail, Chilly Melon Soup and Lightning Chestnut Soup all require less than one minute's mixing-time in the Liquidiser, so they're ideal for a busy hostess. Potted Beef, too, is a joy to make with the Liquidiser. The delicious Pâté of Pork can be made into either a rough or a smooth dish by regulating the running time.

There are enough recipes here for you to serve a different starter at each dinner-party for months to come.

Soups and Starters

From Left to Right.
Florida Cocktail, Potted Beef, Shrimp Avocado
Chilly Melon Soup and Brown Onion Soup

Florida Cocktail

Using Juice Extractor

1 large orange
1 grapefruit

Chill fruit in the refrigerator until required. Cut the fruit in half and extract the juice. Serve immediately.
Serves 2.

Brown Onion Soup

Using Liquidiser or Blender

2-ounce fat
4 large onions (sliced)
2 rashers of bacon (cut up)
1½ pint stock
1 slice of stale bread
Salt and pepper
2-ounce grated cheese

Melt the fat in a saucepan and add the sliced onion and bacon. Cook over a low heat until the onion begins to brown. Add ¼ pint stock and simmer with the lid on for 15 minutes.
Pour into the goblet, add the bread and remaining stock. Blend for 30 seconds on maximum speed.
Pour back into the saucepan, bring to simmering point and season to taste.
Serve with a good sprinkling of grated cheese.
Serves 6.

Potted Beef

Using Liquidiser or Blender

1-pound stewing steak
¼ pint stock
½ onion
1 bay leaf
Salt and pepper to taste
1-ounce butter
1 clove garlic (optional)

Cut the beef into cubes, place in a casserole with the stock, onion, bay leaf and seasonings. Cover the casserole and cook at 350°F/Reg 4 for approximately 2 hours or until tender.
Drain the meat (retaining stock) and remove onion and bay leaf. Add the butter and approximately 4 tablespoons of the stock.
Place half the ingredients in the blender and blend until smooth. Repeat with the other half and add the garlic at this stage if used.
Serve as a sandwich filling or in individual dishes as an hors d'oeuvre.
As this Potted Beef contains no preservative, it will not keep for longer than 2-3 days in a refrigerator.
Serves 4-6.

Tomato Cocktail

Using Liquidiser or Blender

4 tomatoes (quartered)
Piece of onion
1 stick celery
3 sprigs parsley
4 tablespoons water
Squeeze of lemon juice
Dash of Worcestershire sauce
Salt and pepper to taste

Blend all ingredients for 30-40 seconds or until smooth. Strain.
Serves 4.

Shrimp Avocado

Using Liquidiser or Blender

2 fluid ounce mayonnaise
4-ounce shrimps (peeled)
2 ripe avocado pears
1 tablespoon lemon juice
4 lettuce leaves (washed)

Prepare the mayonnaise by following the recipe and instructions on page 46. Measure 2 fluid ounces of mayonnaise into a basin, add the shrimps and mix well. Chill.
Just prior to serving, cut the pears in half (using a stainless steel knife), remove the stone and sprinkle with lemon juice. Fill the hollows with shrimp mixture. Serve on washed lettuce leaf.
Serves 4.

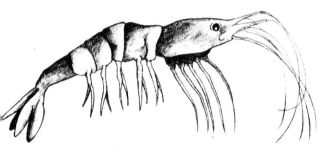

Lightning Chestnut Soup

Using Liquidiser or Blender

½-pound chestnut purée (natural)
1-ounce onion
1-ounce butter
½ pint milk
1 pint white stock (or 1 pint water plus
   chicken stock cube)
½ level teaspoon salt
Shake of pepper
2 tablespoons single cream

Place purée, onion, butter and milk in the goblet and blend for 15 seconds. Add to the stock in a saucepan, season and then bring to the boil. Remove from the heat, adjust seasonings if necessary, stir in cream and serve.
Serves 6.

Chilly Melon Soup

Using Liquidiser or Blender

1-pound melon flesh (1 medium melon)
   (roughly chopped)
1 pint cold milk
1 level teaspoon ground ginger
1-ounce castor sugar

Place all ingredients in the goblet and blend for 10 seconds. Chill thoroughly before serving.
Serves 6.

Egg Mayonnaise

Using Liquidiser or Blender

4 lettuce leaves (washed)
4 hard-boiled eggs
Mayonnaise—see page 46
Cayenne pepper

Place lettuce leaves on individual plates. Slice eggs and arrange on lettuce. Spoon mayonnaise over the egg and sprinkle with cayenne pepper.
Serves 4.

Cheese Meringues

Using Table or Hand Mixer

2 egg whites
2 level tablespoons Parmesan cheese
   (finely grated)
Salt and pepper to season
Fat for deep frying
Parsley for garnish

Whisk the egg whites on maximum speed until they reach meringue peak. Reduce to speed 1, fold in the cheese and seasonings, then switch off immediately.
Heat the fat to 350°F or until a 1 in. cube of bread turns golden brown in 60 seconds. Drop one tablespoon of cheese mixture at a time into the fat and cook until golden brown. Drain well and serve hot.
Serves 2-3.

Above, Pâté of Pork
showing ingredients and finished pâté

Right, Clear Vegetable Soup
Far Right, Herb Potato Soup and
Mushroom and Kidney Soup, *recipe page 16*

Pâté of Pork

Using Liquidiser or Blender

8 bacon rashers
$\frac{1}{2}$-pound pig's liver (cut into pieces)
1-pound lean pork (cut into cubes)
$\frac{1}{4}$ pint cold water
3-ounce breadcrumbs (prepared in blender)
Salt to taste
$\frac{1}{2}$ level teaspoon ground mace
$\frac{1}{2}$ small onion (roughly cut up)
1 level teaspoon rosemary
$\frac{1}{2}$ level teaspoon sage
$\frac{1}{2}$ egg (beaten)

Remove the rind from bacon and press rashers flat with a knife. Line base and sides of a 2-pound loaf tin with the rashers—press firmly.

Place liver, pork and water in a saucepan and simmer over a low heat for 15 minutes.

Prepare breadcrumbs by dropping cubes of bread through the hole in the lid of the blender onto the revolving blades. Tip breadcrumbs into a mixing bowl.

Place salt, mace, onion, rosemary, sage and egg in the blender goblet and blend for 15 seconds. Add to breadcrumbs and stir well.

Drain the meat and then pass each piece through the hole in the lid of the blender onto the revolving blades. When the goblet is approximately one-third full, take out the meat and add to the rest of the ingredients. Continue blending until all the meat has been used.

Mix all ingredients together (this should be of a soft consistency); if slightly too firm, adjust with water from the liver and pork.

Place the pâté mixture in the 2-pound loaf tin, stand in a meat tin of water and cook in a moderate oven 325°F/Reg 3 for 2 hours. The pâté is cooked when the juice pressed out of the centre is no longer pink.

Allow to cool completely before turning out.

Serves 8.

Clear Vegetable Soup

Using Juice Separator

1-pound carrots (well washed)
½-pound potatoes (peeled)
1 medium onion (peeled)
3 tomatoes (washed)
2 sticks of celery (washed)
1 pint water
1 chicken stock cube
1 level teaspoon salt
Shake of pepper

Cut all vegetables into suitably sized pieces for the Juice Separator. Put all vegetables through the attachment (remembering that the basket has to be cleared after each pound of food has been processed). Place the vegetable juices, water, crumbled stock cube and seasonings in a saucepan. Bring to the boil and serve immediately garnished with very thin slices of vegetables.
Serves 4.

Herb Potato Soup

Using Liquidiser or Blender

½-pound potatoes (peeled and sliced)
½ medium onion (sliced)
1 teaspoon mixed dried or fresh herbs ⎱ in a deep
2-ounce butter or margarine ⎰ saucepan
1½ pints white stock, or water plus
   chicken stock cubes
½ level teaspoon salt
¼ level teaspoon pepper
2 tablespoons single cream (optional)

Fry the potatoes, onion and herbs in the butter for approximately 5 minutes. Place the fried food in the goblet with ½ pint stock, and blend for 45 seconds.
Return to the saucepan with the rest of the stock and seasoning, bring to the boil and simmer for 5 minutes. Remove from heat and add cream before serving.
Serves 6.

Mushroom and Kidney Soup

Using Liquidiser or Blender

1-ounce butter ⎱ in a frying pan
¼-pound mushrooms ⎰
1-ounce flour
¼-pound kidney (roughly cut up)
1 pint water
1 stock cube
½ level teaspoon salt
Shake of pepper

Fry the mushrooms in the butter for 3 minutes or until lightly browned. Place fried mushrooms with the rest of the ingredients in the goblet and blend for 45 seconds.
Transfer the soup to a saucepan, bring to the boil slowly and simmer for 10 minutes.
Serve piping hot with croûtons.
Serves 4.

Quick Liver and Bacon Soup

Using Liquidiser or Blender

½-pound liver ⎱ cut up
¼-pound bacon ⎰
1 medium onion (roughly chopped)
1-ounce butter
1-ounce flour
1½ pints water
1 stock cube
1 level teaspoon salt
Shake of pepper
1 dessertspoon lemon juice
2 drops Tabasco sauce

Place liver, bacon, onion, butter and flour with ½ pint water, a stock cube and seasonings in the goblet and blend for 45 seconds. Transfer the blended ingredients to a saucepan, add the other pint of water and simmer for 10 minutes. Remove from the heat, add lemon juice and Tabasco sauce. Adjust seasoning before serving.
Serves 6.

Pâté

Using Liquidiser or Blender

Béchamel Sauce

½ pint milk
1 carrot
½ onion
Bay leaf
Piece of mace
3 peppercorns
1-ounce butter
1-ounce flour

Pâté

¼-pound streaky bacon
½-pound pig's liver
1 clove garlic
½ tin anchovies
2 tablespoons sherry
1 level teaspoon salt
Shake of pepper

Decoration

1 bay leaf
8 rashers bacon

To prepare sauce, bring the milk with the vegetables and spices to the boil, remove from the heat and leave to infuse for 20 minutes; then strain.
Make a roux with the butter and flour, then add the milk.
Prepare a 1-pound loaf tin by placing a bay leaf in the bottom and then by lining it with the 8 bacon rashers.
Remove the rind from the ¼-pound streaky bacon, cut into pieces and fry. Cut the liver into pieces and fry in the bacon fat for a few minutes.
Add the fried bacon and liver, together with the rest of ingredients, to the sauce and stir well. Pour half this mixture into the blender goblet and blend until smooth (20-30 seconds). Repeat with the other half.
Place the pâté mixture in the loaf tin, cover with foil and stand in a meat tin half full of water. Cook at 350°F/Reg 4 for 1-1½ hours or until the juice from the pâté is no longer pink.
Cool completely before serving.
Serves 6.

This section contains such a variety of dishes that you will be able to give your family interesting, quickly prepared dishes for weeks to come. Quick Chicken Surprise is ideal for those occasions when you aren't certain at what time the meal is to be served. Although it takes only minutes to prepare, it can simmer happily in the oven until you are ready to serve it. Bacon in Cider would add an interesting variation to your menus. It is so versatile that it is just as delicious served hot with vegetables as it is cold with salad or in sandwiches.

Anyone who has made Fondue will know that the only tedious part of the preparation is the grating of the cheese, but the Slicer/Shredder can do that for you. Try it for yourself.

Main Dishes

Fondue

Using Slicer-Shredder

1-pound Gruyère cheese
Clove of garlic
½-ounce butter
1 wine-glass dry white wine
1 level teaspoon cornflour
1 liqueur glass Kirsch
Nutmeg and pepper

Shred the cheese on no. 1 drum of the Slicer and Shredder.

Rub the inside of a glazed earthenware casserole or heavy based saucepan with garlic; add the butter and melt over an extremely low heat.

Add the cheese and white wine, stirring gently until cheese has melted. (It is essential to ensure that the cheese mixture does not boil.) Mix the flour and Kirsch and add to the cheese, stirring well. Season with nutmeg and pepper.

Transfer the mixture to a fondue dish and keep hot. Have a basket of cubed French bread ready. According to the Swiss custom each person skewers a piece of bread on a fork and then dips and twirls it two or three times in the cheese.
Serves 4.

Beef in Beer

Using Liquidiser or Blender

1-ounce butter
1½-pound best stewing ⎤
  steak (cubed)   ⎬ tossed together
1-ounce plain flour ⎦
½ pint brown ale
Salt and pepper
3 onions (roughly chopped)
1 teaspoon vinegar
½ level teaspoon mace
1 teaspoon brown sugar
1 teaspoon Worcestershire sauce

Melt the butter in a saucepan and gently fry the meat for 4-5 minutes. Place the meat in the bottom of a large casserole. Put the rest of the ingredients in the goblet and blend for 30-45 seconds. Pour the contents of the goblet over the beef, cover and bake for 2 hours in a pre-heated oven, 350°F/Reg 4.
Serves 4-6.

Cheese Pudding

Using Slicer-Shredder and Liquidiser

5-ounce Cheddar cheese (grated on no. 1 drum)
3-ounce white bread (crumbed in liquidiser)
3 eggs
Pinch of salt, pepper and mustard
1 level teaspoon mixed herbs
¾ pint milk   ⎤
1-ounce butter ⎬ in a saucepan
Parsley for garnish

Place 4-ounce of grated cheese, breadcrumbs, eggs, seasonings and herbs in the bowl. Mix on speed 3 for 1-1½ minutes. Pour the warm milk and butter over this and continue to mix for a further 30 seconds. Pour the mixture into a greased pie dish and then sprinkle the rest of the cheese on top. Bake at 375°F/Reg 5 for 30-40 minutes or until firm. Garnish with parsley and serve piping hot.
Note. A little chopped ham makes this even more tasty!
Serves 4.

20

Cheese Potato Pie

Using Slicer-Shredder and Potato Peeler

1-pound potatoes (peeled)
4-ounce cheese
1 small onion
1-ounce butter
Salt and pepper

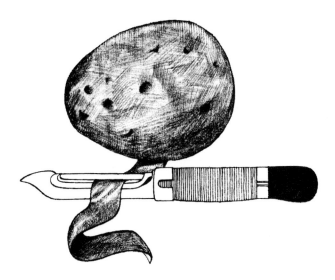

Slice the potatoes on the Slicer and Shredder. Wash well in cold water, then drain. Grate the cheese and onion.

Use a little of the butter to grease the inside of an ovenproof dish. Arrange an overlapping layer of the potato slices in the bottom. Sprinkle with grated cheese, onion, salt and pepper. Cover with another layer of potato, then dot with butter. Continue in this way alternating layers of potato and cheese, finishing with a cheese layer. Cover the dish and bake at 350°F/Reg 4 for 30 minutes, then remove the cover and bake for a further 30 minutes at the same temperature; test the potatoes with a sharp knife to ensure that they are cooked. Serve immediately.

This is delicious served with fresh cooked sausages.

Serves 3.

Lamb and Tomato Casserole

Using Liquidiser or Blender

2-pound best end of neck of lamb cutlets
2 onions (roughly chopped)
½ green pepper (roughly chopped)
2-ounce mushrooms
1 tablespoon parsley
1 large can tomatoes
1 teaspoon Worcestershire sauce

Wipe lamb, trim off surplus fat and fry gently for approximately 5 minutes. Place the cutlets in the bottom of an ovenproof dish. Place the rest of the ingredients in the goblet and blend for 30 seconds. Pour the contents of the goblet onto the cutlets and cover. Bake at 375°F/Reg 5 for 1 hour.

Serves 6.

Baked Onions with White Sauce

Using Liquidiser or Blender

1-pound medium-sized onions
 (peeled but left whole)
½-ounce butter
1 tablespoon water

Sauce

1-ounce butter or margarine
1-ounce plain flour
½ pint milk
Salt and pepper to taste
Sprinkling of mustard powder

Place the onions in a saucepan, cover with cold water and bring to the boil. Simmer for 1 minute. Drain and place in a casserole dish together with butter and water. Cover and cook in 350°F/Reg 4 oven for 45-60 minutes or until tender.

Meanwhile, place all the ingredients for the sauce in the blender, switch on to maximum speed and blend for 20-30 seconds. Pour into a saucepan and cook over a low heat stirring all the time, until thick (2-3 minutes).

When the onions are cooked and ready to serve, remove the lid from the casserole, drain off any liquid and pour the sauce over.

Serves 4.

Bacon in Cider

Using Liquidiser or Blender

2-pound bacon joint
1 medium onion (roughly chopped)
2 carrots (roughly chopped)
15 fluid ounce cider
Salt and pepper to season
1 bay leaf

Trim any surplus fat from the bacon joint. If possible, soak the joint overnight in cold water and then discard this water. Transfer the joint to a deep casserole.

Place the onion, carrots, cider and seasoning in the goblet and blend for approximately 30 seconds. Pour the contents of the goblet over the joint. Place the bay leaf on the bacon, cover the casserole and cook at 350°F/Reg 4 for $1\frac{3}{4}$-2 hours. Eat either hot or cold.

Note. The cider and vegetables can be thickened with flour to make a sauce.

Serves 4-6.

Coq au Vin

Using Liquidiser or Blender

4 joints of roasting chicken ⎫
2 rounded tablespoons flour ⎬ together
1 level teaspoon salt ⎭
2-ounce butter (melted)
1 large onion (roughly chopped)
1 clove garlic (roughly chopped)
$\frac{1}{4}$ green pepper (roughly sliced)
$\frac{1}{4}$-pound lean bacon (cut up)
2 tablespoons parsley (chopped)
$\frac{1}{2}$ pint red wine
Bay leaf
$\frac{1}{4}$-pound mushrooms

Toss the joints in the seasoned flour and then fry gently in a large saucepan. Fry until crisp and golden on both sides. Remove to a plate. Add chopped onion, garlic, pepper, bacon and mushroom stalks to the remaining butter, fry gently until pale gold. Place the ingredients from the saucepan together with the parsley and wine in the goblet and blend for 30 seconds.

Return the chicken pieces to the saucepan, add the blended ingredients and bay leaf and simmer for 1 hour. Add the mushrooms and simmer for a further 15 minutes.

Serves 4.

Spanish Pancakes and
Rarebit with a Difference, *recipe page 26*

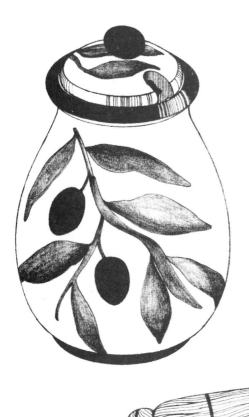

Spanish Pancakes

Using Liquidiser or Blender

Batter

1 egg
½ pint milk
4-ounce plain flour
½ level teaspoon salt

Filling

1-ounce butter
1 large Spanish onion (sliced)
4 tomatoes (sliced)
½ green pepper (sliced)
2-ounce mushrooms (sliced)
Salt and pepper
2-ounce anchovy fillets (drained)
Black olives (quartered)

Place all the batter ingredients in the goblet and blend until smooth—approximately 15 seconds. Put the mixture to one side.

Melt the butter in a large saucepan, add the onion, tomato, pepper, mushrooms and seasoning, cover the pan, turn the heat very low and allow the vegetables to cook slowly until they become tender—10-15 minutes approximately. If they are very moist, then pour off the surplus liquid before adding the anchovies and olives.

Whilst the filling is cooking, prepare the pancakes in the usual way, keeping them warm in the oven. A piece of greaseproof paper can be placed between each one to keep them separate. Just prior to serving, place the filling on each pancake, roll up and serve immediately.

Serves 4.

Tomato Mould

Using Liquidiser or Blender

1 small tin tomatoes
1 small tin tomato juice
$\frac{1}{2}$ level teaspoon salt
Dash of pepper
$\frac{1}{4}$ of a medium onion (roughly chopped)
1-ounce gelatine
Dash of Worcestershire sauce

Cheese Filling

4-ounce cottage cheese
2 tablespoons mayonnaise
2-ounce blue cheese

Simmer the tomatoes, juice, seasoning and onion for 10 minutes. Allow to cool, then pour into the goblet. Soften the gelatine in a little water and add to the tomato mixture, together with a dash of Worcestershire sauce. Blend for 20 seconds. Pour through a strainer to remove seeds and then into either a pint or individual moulds. Leave to become cold, then place in the refrigerator to set. Place the cottage cheese and mayonnaise in the goblet and blend for 15 seconds. Remove the centre cap of the lid and drop the blue cheese in through this onto the revolving blades. The motor may have to be stopped to push the food back onto the blades.

Serve the cheese filling with the mould.

Sweet Corn Soufflé

Using Whisk and Can Opener

$1\frac{1}{2}$-ounce butter
$1\frac{1}{2}$-ounce flour
$\frac{1}{2}$ pint milk
Seasoning
Small tin sweet corn (drained)
1 tablespoon green pepper (finely chopped)
3 egg yolks
4 egg whites

Melt butter in saucepan, add flour and cook gently for 1 minute. Remove from heat, add milk and seasoning and stir until smooth. Cook sauce stirring constantly for 3 minutes. Allow to cool slightly.

Mix in sweet corn and green pepper, then beat in egg yolks one at a time. Whisk egg whites on maximum speed until stiff.

Turn the sweet corn mixture into the egg whites, fold carefully together with a large metal spoon. Pour into a 1-pint prepared soufflé dish. Bake in a moderate oven (375°F/Reg 5) for 45-50 minutes. Serve immediately.

Serves 4.

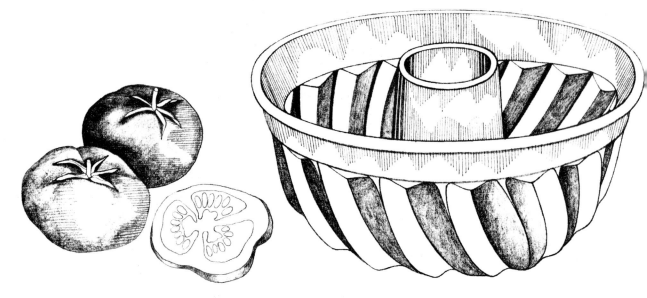

Puffed Omelette

Using Table or Hand Mixer

2 eggs (separated)
Salt and pepper
Oil for frying

Whisk the egg whites until they stand in peaks, then fold in the yolks plus seasoning. Heat the oil in an omelette pan, add the mixture and cook until it is lightly browned on the underside. Transfer the partly cooked omelette to a medium grill and cook until it is just firm. Place the cooked filling on one half, fold over the other and serve immediately.

Suggested Fillings

Tomato—Skin and slice 2 tomatoes. Fry in a tablespoon of oil, then add ½ level teaspoon of mixed herbs. Fold into cooked omelette.

Spanish—Skin and slice 2 tomatoes and 2 onions. Fry these slices in a tablespoon oil until tender but not brown. Add 1 tablespoon of tarragon vinegar and a shake of cayenne pepper. Fold into cooked omelette.

Mushroom—Cut 2-ounce mushrooms into pieces and sauté in 1-ounce butter. Fold into cooked omelette.

Kidney—Cut up a sheep's kidney and sauté in ½-ounce butter for 15-20 minutes or until tender. Just before the cooking is complete, add 1 teaspoon of paprika pepper. Fold into cooked omelette.
Serves 1.

Quick Chicken Surprise

Using Table Mixer

1 tin condensed soup (chicken or mushroom)
2 tablespoons mayonnaise—see page 46
Salt and pepper
3-ounce celery (chopped)
2-ounce nuts (chopped in liquidiser)
1-pound cold chicken meat (off the bone)
2 hard-boiled eggs (quartered)
2 packets crisps (crushed)

Place the soup, mayonnaise and seasoning in the bowl and mix on speed 3 until they are combined. Reduce to speed 1 and stir in the celery, nuts and chicken.
Place the pieces of egg in the bottom of a casserole and spread the chicken mixture on top. Cover with a liberal amount of crushed crisps. Cook at 350°F/Reg 4 for 30-40 minutes or until piping hot.
Serves 4.

Cabbage Salad

Using Table Mixer with Slicer-Shredder and Liquidiser

2-ounce raisins ⎫ together in
Juice of 1 lemon ⎭ a bowl
½ medium cabbage
½ medium onion (sliced)
½ green pepper (sliced)
6 tablespoons salad oil
2 tablespoons tarragon vinegar
2 level teaspoons sugar
½ level teaspoon salt
Pepper

Soak the raisins in the lemon juice for several hours. Drain but retain the juice. Shred the cabbage on the Slicer and Shredder. Place this in a salad bowl together with the onion, green pepper and raisins.
Place remaining lemon juice, oil, vinegar, sugar, salt and pepper in the goblet and blend for 10 seconds. Pour over salad, tossing lightly.
Serves 4.

Jan 29/85 Excellent

Rarebit with a Difference

Using Liquidiser or Blender *Use as an appetizer*

1 medium tomato (cut into 4)
½ medium onion (roughly chopped)
2 tablespoons milk
8-ounce cheese (cubed)
4 slices toast

Can add dragons.

Place tomato, onion and milk in the goblet and blend for 15 seconds. Remove the centre cap from the lid and drop the cubes of cheese onto the blades. The motor may have to be stopped to push the mixture back onto the blades. If the mixture is not moist enough, then add a little more milk. Spread on toast and cook under a hot grill. Garnish with watercress or parsley.
Serves 4.

Haddock with Almonds and Lemon

Using Liquidiser or Blender

3-ounce breadcrumbs (made in goblet)
1 clove of garlic
1 large sprig of parsley
1-ounce almonds
2-ounce butter (melted)
Finely grated rind and juice of 1 large lemon
4 haddock steaks
Salt and pepper

Prepare the breadcrumbs by dropping cubes of bread onto the revolving blades. Empty all the crumbs out, then chop up finely the garlic, parsley and almonds in the same way. Mix the breadcrumbs, almonds, parsley and garlic together, stir in the lemon rind, and then add this to the melted butter, frying gently until all the butter has been absorbed. Place the haddock steaks in the bottom of a well-buttered casserole dish, sprinkle lemon juice over and then top with the fried crumbs. Bake uncovered at 350°F/Reg 4 for approximately 30 minutes or until the fish flakes.
Serves 4.

Sausage Casserole

Using Liquidiser or Blender

1-pound chipolata sausages
1 medium onion (sliced)
3 rashers of bacon (trimmed and cut into pieces)
1 level tablespoon flour
2 teaspoons soy sauce
¼ pint milk
3 sprigs of parsley
Salt and pepper
1 small tin sweetcorn
3 slices of bread (crumbed in liquidiser)
1-ounce butter

Fry the sausages until lightly browned. Remove from pan and place in the bottom of an ovenproof dish. Pour most of the fat off, then fry the onion and bacon for a few minutes.

Place the flour, sauce, milk, parsley and seasoning in the goblet and blend for 15 seconds. Pour the contents of the goblet into the frying pan with the onion and bacon—cook this over a low heat stirring all the time, until the mixture thickens. Stir in the drained corn. Pour the corn mixture over the sausages, then sprinkle liberally with the breadcrumbs. Dot with butter. Bake uncovered at 350°F/Reg 4 for 30-40 minutes.
Serves 4.

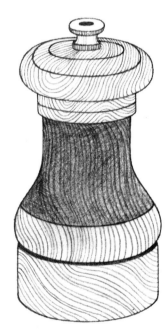

Lancashire Hot Pot

Using Slicer-Shredder

$1\frac{1}{2}$-pounds best end of neck of lamb
1-pound potatoes (peeled)
$\frac{1}{2}$-pound carrots (peeled)
$\frac{1}{2}$-pound onions (peeled)
$\frac{1}{4}$ pint stock
Pinch of thyme } mixed together
Salt and pepper
Bay leaf

Cut the best end of neck into cutlets. Using the no. 3 drum of the Slicer and Shredder slice the potatoes, carrots and onions. Arrange the meat and vegetables in alternate layers in a casserole. Pour the mixture of stock, thyme and seasoning over the casserole. Place the bay leaf on top. Cover and cook at 300°F/Reg 2 for 2-2$\frac{1}{2}$ hours.
Serves 4.

Mushroom Soufflé

Using Liquidiser or Blender

$\frac{1}{2}$-pound mushrooms (sliced)
2-ounce butter
1 teaspoon grated onion
1-ounce flour
8 fluid ounce milk or single cream
4 eggs (separated)
Seasoning to taste

Melt $\frac{1}{2}$-ounce butter in a saucepan and sauté the mushrooms until barely cooked. Stir in the grated onion and put to one side. Place 1$\frac{1}{2}$-ounce butter, the flour and milk in the goblet and blend for 30 seconds. Pour the contents of the goblet into a saucepan, bring to the boil stirring continuously. Beat in the egg yolks one at a time, then fold in the mushroom mixture.
Whisk the egg whites until they hold soft peaks and then fold into the sauce. Season. Transfer the mixture to a greased 2 pint soufflé dish. Cook at 375°F/Reg 5 for 45-50 minutes or until firm and golden brown.
Serves 3 or 4.

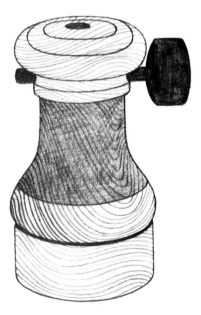

With such sweets as Strawberry Ice-Cream Pudding, Meringue Nests and Milanese Soufflé, one soon discovers that using a foodmixer for whisking egg-whites and cream is a great boon when preparing something special to round off a meal. Sweets that were at one time impractical because of the time involved can now come easily within your scope. Or if your family's taste is simple, and they prefer to finish their meals with fruit, use the cream-maker to produce the accompanying cream or ice-cream at a fraction of the cost.

Hot and Cold Sweets

Meringue Nest with Fruit

Fruit Cobbler

Using Table or Hand Mixer

1½-pound tinned or stewed fruit

Cobbler Topping

8-ounce self raising flour (sieved)
Pinch of salt
2-ounce margarine
1-ounce castor sugar
4 fluid ounce milk

Place the fruit in a pie dish.
Mix the sieved flour and salt with the margarine on a low speed until it resembles breadcrumbs. Add the sugar then the milk, mixing just enough to incorporate it to form a soft dough.
Roll the dough to ¾ in. thickness and cut into rounds with a 2 in. fluted cutter. Arrange these on top of the fruit, brush them with milk and sprinkle with castor sugar. Bake at 400°F/Reg 6 for 20-25 minutes.

Crumble with a Difference

Using Table or Hand Mixer

1½-pound of any cooked fruit (sweetened to taste),
 or 1 large can fruit (drained)
4-ounce plain flour
2-ounce margarine
2-ounce demerara sugar
1-ounce nuts (roughly chopped)
1-ounce polka dots

Place the cooked fruit in the bottom of a heat-resistant pie dish.
Place the flour and margarine in the bowl and mix on a low speed until the mixture resembles bread-crumbs. Using the same speed stir in the sugar, nuts and polka dots. Sprinkle the mixture on top of the fruit and cook at 350°F/Reg 4 for approximately 30 minutes.
Serves 4.

Plum Meringue Pie

Using Table or Hand Mixer

4-ounce plain flour
Pinch of salt
2-ounce margarine (or 1-ounce margarine plus
 1-ounce lard) mixed together
1 egg yolk
2 teaspoons cold water

Filling

1 large tin of plums
1 egg yolk

Meringue

2 egg whites
4½-ounce icing sugar (sieved)

Sieve the flour and salt together in the bowl. Add the fat and mix on a low speed until the mixture resembles breadcrumbs. Add sufficient of the egg yolk and water mixture to form a pastry dough. Switch the machine off immediately the liquid is incorporated.
Line a 7 in. flan ring with the pastry and bake blind at 425°F/Reg 7 for 20 minutes approximately, or until the case is firm and lightly browned.
Remove the stones from the plums and beat in the egg yolk. Spread this fruit purée in the bottom of the pastry case.
Place the egg whites in the bowl, add the sugar and beat on maximum speed, until the mixture is firm and stands in peaks.
If a hand mixer is used, then this mixture should be whisked over hot water.
Pile the meringue on top of the plums, then bake at 325°F/Reg 3 for approximately 30 minutes.
Note. A pound of any fruit purée could be used in this recipe, instead of plums.

32

Vanilla Ice-Cream

Using Cream Maker

4-ounce unsalted butter
8 fluid ounce milk
3-ounce sugar
1 level teaspoon vanilla essence

Place the butter, milk and sugar in a saucepan, heat slowly until the butter has melted and allow just to come to the boil. Ensure that the 'grooved hand nut' is screwed home before pouring mixture into the container of the main body. Place a suitable receptacle beneath the valve to collect the ice-cream liquid.

Switch on the mixer to speed 3 and allow to run until all the mixture has passed through. Stir in the vanilla essence and then pour into the freezer tray. Freeze until there is a solid film ½ in. around the edge of the tray.

Scrape the half-frozen mixture into the bowl and whisk on speed 4 for approximately 30 seconds. Turn back into the ice tray. Freeze until firm.

To obtain flavoured ice-cream, make up the recipe above (omitting vanilla essence) as far as the stage where the ice-cream leaves the Cream Maker. Pour this liquid into the goblet of the liquidiser, and the flavouring and blend on maximum speed for 30 seconds. Transfer this to the freezer tray and freeze, following the instructions for Vanilla Ice-Cream.

Serves 3-4.

Strawberry or Raspberry Ice-Cream
Add 4-ounce fresh or frozen fruit.

Orange, Lemon or Lime
Add the grated rind of the fruit and 2 tablespoons of juice.

Chocolate
Add 2-ounce chocolate (broken into squares).

Coffee
Add 1 teaspoon coffee essence.

Peppermint
Add 3 drops of peppermint essence and a little green colouring.

C

Ginger
Add ½ teaspoon liquid ginger flavouring and 1-ounce chopped crystallised ginger (optional).

Strawberry Ice-Cream Pudding

Using Liquidiser or Blender

½ pint milk
4 level tablespoons cornflour
6-ounce castor sugar
1-pound fresh ripe or frozen strawberries
½ pint cream (stiffly whipped)
2 tablespoons blackcurrant jelly

Decoration

A little whipped cream
Fresh strawberries

Mix a third of the milk with the cornflour, bring the rest of the milk to the boil and pour it onto the mixture. Return to the saucepan, bring to the boil and allow to boil gently for 2 minutes, stirring all the time. Add the sugar, stir well and then leave to cool.

Pour all ingredients (including cool sauce) into the goblet and blend until smooth. Pour into a 1½ pint ring mould and freeze for several hours. Dip the mould in cold water to ease out the sweet. Unmould onto a plate. Return to freezing compartment to become completely firm again. Fill centre with fresh strawberries and decorate with cream.

Serves 6.

Chocolate Lime Splice

Using Table or Hand Mixer

Crust

1-ounce margarine
3-ounce plain chocolate
1 rounded tablespoon golden syrup
2-ounce cornflakes

} in a basin

Filling

2 level teaspoons cornflour
2½ fluid ounce concentrated
   lime cordial
2½ fluid ounce water
2-ounce castor sugar

} in a small saucepan

1 egg (whisked until frothy)
½ pint double cream (whisked)

Decoration

Chocolate curls

Grease pie dish

Melt margarine, chocolate and syrup in a bowl over hot water and mix in the cornflakes. Press the mixture into the bottom and sides of an 8 in. pie dish to form a crust. Leave in a cool place to set.

Filling

Mix the cornflour with the liquid until smooth in a small saucepan. Bring to the boil and cook for a few minutes until it is thickened. Beat in the egg and sugar on speed 3—then allow the mixture to cool. Fold in the whisked cream on speed 1. Add a drop of green colouring at this stage if required. Pour into the cornflake crust. Chill and decorate with chocolate curls.
Serves 6.

Sun Splash Cheese Cake

Using Liquidiser or Blender

5-ounce digestive biscuits (crumbled in liquidiser)
1½-ounce butter or margarine (melted)
1-ounce gelatine
½ pint warmed milk

} together

8-ounce cream cheese (cut into pieces)
1-ounce castor sugar
Rind of 1 orange
2 eggs (separated)

Decoration

Segments from 1 orange
½ glacé cherry

Mix together the digestive biscuits and melted butter until the crumbs are very moist. Press these firmly into the bottom of a lightly greased 6 in. loose-bottomed cake tin. Place the dissolved gelatine, cream cheese, castor sugar, orange rind and egg yolks in the liquidiser goblet and blend for 45 seconds or until smooth. Whisk the egg whites to meringue peak and then gently fold in the contents of the liquidiser. Pour this onto the biscuit-crumb base and place in refrigerator to become quite firm.

Decoration

Pare all the skin and pith off the orange and then cut out the flesh in whole segments using a sharp knife. Arrange to form a flower with the cherry as the centre.
Serves 4-6.

Right, Sun Splash Cheese Cake
Lemon Sorbet and Chocolate Lime Splice

Meringue Nests

Using Table or Hand Mixer

4 egg whites
9-ounce icing sugar (sieved)

Filling

¼ pint double cream
Fresh or tinned fruit

Decoration

3-ounce grated chocolate

Brush the baking trays lightly with oil. Place the egg whites and icing sugar in the bowl and whisk on maximum speed until it stands in peaks. Alternatively, using the Hand Mixer, whisk on maximum speed over a saucepan of boiling water for about 5 minutes.

Fit a large piping bag with a ½ in. star nozzle and half fill with meringue mixture. Pipe nests onto prepared trays in the following way:

Holding the bag upright, pipe the base in a 'catherine wheel' formation until 3 in. in diameter. Pull the pipe away and then pipe a circle on top of the outside edge to make the sides.

Bake the meringues in a very slow oven 200°F/Reg ¼ until they are firm and crisp.

Filling

Whisk the cream until stiff and fill each nest with fruit and cream. Sprinkle with grated chocolate. **Makes 8 nests.**

Syllabub

Using Table or Hand Mixer

1 egg white
3 tablespoons double cream
1 dessertspoon brandy
1 dessertspoon sweet sherry

Whisk egg white until it holds its shape. Then whisk the cream until it forms soft peaks. Mix gently together with the sherry and brandy. Serve with sweetened fresh fruit.

Thick Cream

Using Cream Maker

3 or 4 ounce unsalted butter
4 fluid ounce milk

Place the milk and butter in a saucepan, heat slowly until the butter has melted and then allow to just come to the boil.

Ensure that the 'grooved hand nut' is screwed home before pouring mixture into the container of the main body.

Place a suitable receptacle beneath the valve to collect the cream. Switch on the mixer to speed 3 and allow to run until all the mixture has passed through.

After cooling in a refrigerator, lightly mix with a fork before serving. If the cream is required for whipping, it should be cooled in a refrigerator for at least 8 hours and then whisked to piping consistency using the Chef whisk.

Pouring Cream

Reduce the unsalted butter content to 2-ounce and make in exactly the same way as for Thick Cream. To ensure success, always use the Cream Maker with warm ingredients.

Note. Reconstituted milk can be used in place of fresh milk, if made to the manufacturer's instructions.

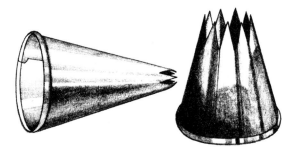

Caramel Fruit

Using Table or Hand Mixer

1-pound fresh fruit or
  1-pound tinned fruit (drained)
½ pint double cream
3-4-ounce soft brown sugar

Prepare the fruit, i.e. strawberries washed and hulled, apples sliced, and cover the base of a shallow, heat-resistant dish. Whip the cream on maximum speed until it holds its shape. Spread the cream over the fruit and chill thoroughly.
Just prior to serving, sprinkle thickly with the sugar and place under a very hot grill until the sugar melts and bubbles (about 3 minutes).
Serve immediately.
For an extra special sweet, try fresh peaches halved with a spoonful of brandy over each.
Serves 4.

Lemon Sorbet

Using Blender or Table Mixer with Liquidiser and Whisk

18 fluid ounce water
4-ounce castor sugar
1 large lemon (quartered)
1 egg white

Heat water and sugar gently in a heavy-based saucepan until the sugar is dissolved. Bring to the boil vigorously for 5-6 minutes, leave to cool.
Pour syrup into the goblet, add the lemon and blend for 4 seconds approximately.
Strain the liquid into an ice tray and freeze until just firm. Whisk the egg white until it forms soft peaks. Scoop the frozen mixture from the tray, beat the egg white into this, then return to the ice tray and refreeze before serving.
Serves 4-6.

American Holiday Pudding

Using Table Mixer and Slicer-Shredder

4-ounce self-raising flour
1 level teaspoon baking powder } sieved
½ level teaspoon nutmeg } together
1 level teaspoon cinnamon
1 rounded tablespoon black treacle
2-ounce brown sugar
4-ounce carrot (shredded on no. 2 drum
  of Slicer-Shredder)
4-ounce apple (shredded on no. 2 drum
  of Slicer-Shredder)
4-ounce seedless raisins
2-ounce breadcrumbs (made in liquidiser)
8-ounce suet (grated)
2 eggs

Sieve dry ingredients into the bowl. Add all other ingredients and mix thoroughly on medium speed. Place mixture in a 1½ pint greased pudding basin. Cover and steam for 2½ hours.
Try this sweet with Sabayon Sauce (see page 47).
Serves 4-6.

Norwegian Cream

Using Slicer-Shredder and Whisk

2 rounded tablespoons apricot jam
2 eggs
1-ounce castor sugar
¾ pint milk
¼ pint double cream (whipped)
2-ounce chocolate (grated on Slicer-Shredder)
1-ounce walnuts (grated on Slicer-Shredder)

Spread the jam on the base of an ovenproof pie dish (7 in. or 8 in. in diameter). Whisk together the eggs, sugar and milk and pour this over the jam. Stand the ovenproof dish in a meat tin containing water and bake at 350°F/Reg 4 for 35-40 minutes or until the egg mixture is set. Allow to cool slightly and then refrigerate until completely cold.
Just prior to serving, spread with whipped cream and sprinkle liberally with chocolate and nuts.
Serves 4-6.

Fruit Upside-Down Pudding

Using Table or Hand Mixer

1 tablespoon golden syrup
1 small tin of fruit
4-ounce margarine
4-ounce castor sugar
4-ounce self raising flour
2 eggs

Grease an 8 in. deep-sided sandwich tin lightly and spread a tablespoon of golden syrup in the base. Drain the juice from the fruit (reserve this for Syrup Sauce, page 45) and arrange fruit on top of the syrup. Make a Victoria Sponge following the instructions on page 102 and place the cake mixture on top of the fruit. Bake at 325°F/Reg 3 for 35-40 minutes. Turn out immediately onto a warmed plate and serve with Syrup Sauce.
Serves 4-6.

Baked Chocolate Soufflé

Using Table or Hand Mixer

4-ounce plain chocolate ⎫
1 teaspoon instant coffee ⎬ in a small saucepan
¼ pint milk ⎭
1-ounce butter
1-ounce flour
1-ounce sugar
3-4 eggs (separated)
1 teaspoon rum
or 1 teaspoon vanilla essence

Dissolve the chocolate and coffee in the milk and allow to cool. Melt the butter in a saucepan, add the flour and cook the roux for a few minutes before adding the milk mixture and sugar. Bring to the boil stirring continuously, then beat the egg yolk into the mixture.
Whisk egg whites stiffly, then fold into the chocolate mixture with the rum or the vanilla essence and use speed 1 for the folding in. Pour into a 1 pint soufflé dish and cook for 30-35 minutes at 375°F/Reg 5.
Serve immediately with a chocolate sauce (page 48).
Serves 4-6.

Crème Brulée

Using Hand Mixer

½ pint double cream
2 egg yolks
2 teaspoons castor sugar
2 tablespoons demerara sugar

Place cream, egg yolks and castor sugar in a basin standing over a saucepan of hot water. Whisk over the heat, using Chefette or Mini, until the mixture thickens. Do not boil. Pour into serving bowls, cool and cover with demerara sugar, and brown slowly under the grill until sugar has melted.
Serves 4.

Chocolate Almond Pudding

Using Table Mixer

4-ounce castor sugar
4-ounce pre-creamed margarine
2 eggs
6-ounce self raising flour ⎫
1 level teaspoon baking powder ⎬ sieved together
2 level tablespoons cocoa ⎫ mixed to a paste and
2 tablespoons boiling water ⎬ allowed to cool
½ teaspoon vanilla essence
2-ounce blanched almonds (chopped in liquidiser)

Place all ingredients in the bowl and beat on speed 4 for 2 minutes or until the mixture is perfectly smooth. Transfer the mixture to a greased pudding basin and cover with foil. Steam for 2 hours. Serve immediately with either custard or chocolate sauce.
Serves 4-6.

Opposite, Fruit Upside-Down Pudding

Above, Chocolate Almond Pudding
and Peach Branaska

Peach Branaska

Using Table or Hand Mixer

Sponge

2 standard eggs
2-ounce castor sugar
2-ounce self raising flour
3-4 small peaches (stoned, skinned and sliced)
3 tablespoons brandy
1 family-sized block of ice-cream (sliced)

Almond Meringue

3 egg whites
6-ounce castor sugar
3-ounce blanched almonds (chopped in liquidiser)

Chocolate Sauce

1-ounce plain chocolate ⎫
¼-ounce butter ⎬ in a bowl

Place eggs and sugar in the warmed bowl and whisk on maximum speed until mixture is thick and light coloured (the whisk should leave a trail when lifted out of the mixture). Fold in the flour carefully, using a spatula. Pour the mixture into an 8 in. greased flan tin. Bake for 25 minutes at 350°F/Reg 4. Cool on a wire rack.
Soak peaches in brandy for about ½ hour before required. Whisk egg whites on maximum speed until slightly foamy. Add half the sugar and continue whisking until meringue stands in peaks. Add remaining sugar and whisk for a few seconds to incorporate. Remove the whisk from the hub and use it manually to add the chopped almonds. Place the brandy-soaked peaches in the sponge flan and allow remaining brandy to soak into the sponge. Place slices of ice-cream on top and spread the almond meringue over the whole dish. It must completely cover the ice-cream to prevent it from melting. Flash bake for 2-3 minutes at 475°F/Reg 9.
Melt chocolate and butter together in a bowl over hot water. Drizzle chocolate sauce over surface and serve immediately.
Serves 6-8.

Coconut Flan

Using Table Mixer, Liquidiser, Slicer-Shredder

Ginger Crumb Crust

8-ounce gingernut biscuits (crumbed in liquidiser)
4-ounce butter or margarine (melted)

Filling

½ pint milk
2 level tablespoons cornflour
2-ounce sugar
2 eggs (separated)
4-ounce dessicated coconut
¼ teaspoon vanilla essence

Decoration

1-ounce brazil nuts (sliced on no. 4 drum
  of Slicer-Shredder)

Mix together the melted fat and the biscuit crumbs. Press firmly into a 7 in. fluted flan ring. Leave in a refrigerator to set.
Place the milk, cornflour and sugar in the liquidiser goblet and blend for 30 seconds. Pour the contents of the goblet into a saucepan and bring to the boil. Boil for 2 minutes, stirring all the time. Beat in the egg yolks, coconut and vanilla essence.
Whisk the egg whites until stiff and then fold into the coconut mixture. Pour into the flan case and decorate with flaked brazil nuts.
Chill thoroughly before serving.

Milanese Soufflé

Using Hand Mixer

½-ounce gelatine ⎫
2 fluid ounce water ⎬ in a small basin
2-ounce castor sugar
2 lemons
2 eggs (separated)
½ pint double cream

Decoration

Chopped almonds or ratafias (crushed)
  and cream

Dissolve the gelatine in the water over a small saucepan of hot water. Place in a heat-resistant bowl the sugar, grated lemon rind, juice and egg yolks. Whisk this over a saucepan of boiling water on maximum speed or until the mixture is light and fluffy. Remove from the heat and continue whisking for a few minutes.
Whisk the cream until thick and the egg whites until they hold soft peaks. Fold the cream into the lemon mixture, then the egg whites and gelatine. Pour into a 1 pint soufflé dish banded with a collar of oiled greaseproof paper. Leave in the refrigerator to set.
Remove the greaseproof paper and cover the sides with chopped almonds or crushed ratafia biscuits. Decorate with piped cream.
Serves 4-6.

It has never been so easy to make delicious smooth sauces as it is now with the help of your food mixer. The Liquidiser attachment, in particular, enables you to forget that lumpy sauces ever existed, and gives you the confidence to go on and create more exotic accompaniments than you ever thought possible. Strawberry Whip, Orange Honey Butter and Cold Chocolate Sauce all take only minutes to prepare but can turn something as simple as ice-cream into an exciting new dish.

Mayonnaise, made in the liquidiser, couldn't be simpler. But don't stop there—transform it into Tartare Sauce to enhance your fish dishes.

Sauces

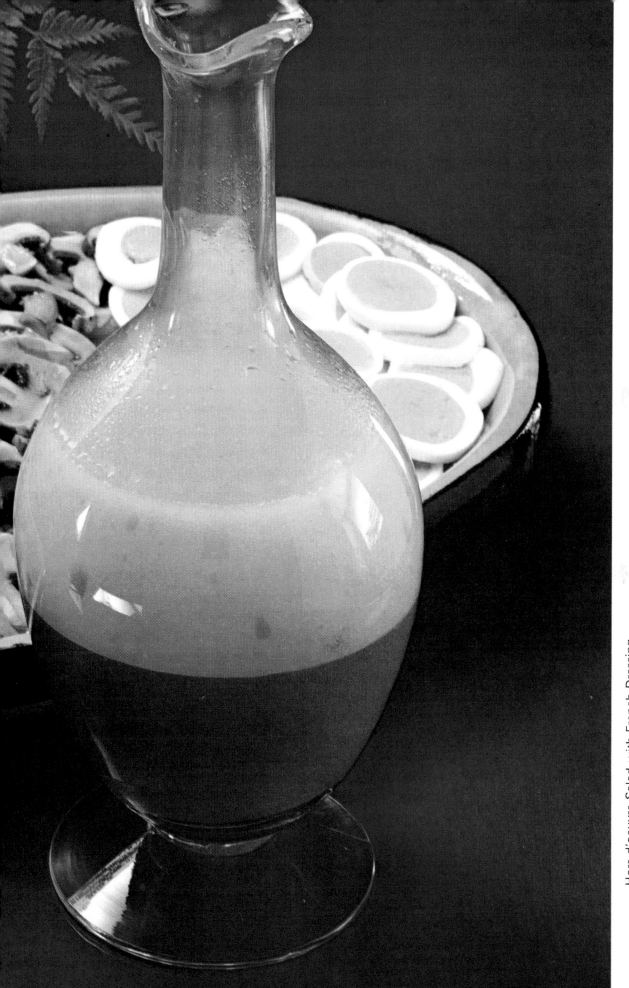

Hors d'oeuvre Salad with French Dressing

Bread Sauce

Using Liquidiser or Blender

3-ounce bread (crumbed in liquidiser)
1 small onion
2 cloves
½ pint milk
1-ounce butter
Salt and pepper

Prepare the breadcrumbs by feeding them through the hole in the lid onto the revolving blades.
Stick the cloves into the onion and place in the saucepan with the milk. Bring the milk to the boil, then set the saucepan aside in a warm place for about 30 minutes.
Remove the onion, pour the milk into the goblet with the bread. Add the butter and seasoning. Blend for 30 seconds. Pour back into the saucepan and reheat before serving.

Corn Relish

Using Liquidiser or Blender

8 fluid ounce vinegar
1½ level teaspoons salt
3-ounce soft brown sugar
1 green or red pepper (seeded and roughly cut)
½ large onion (roughly chopped)
1 large (11-ounce) can of sweet corn

Put all ingredients except corn in the goblet. Blend until the vegetables are finely cut (approximately 30 seconds). Pour into a saucepan and add corn. Cook slowly for 20 minutes. Pour into hot, sterilised jars and seal at once.
This relish is delicious with cold meats and salads.

Curry Sauce

Using Liquidiser or Blender

2 onions (roughly chopped)
2-ounce butter or margarine
1 level tablespoon curry powder
1½ level tablespoons plain flour
1 cooking apple (peeled and roughly cut up)
½ pint stock
Juice of half a lemon
2 level tablespoons brown sugar
Salt and pepper

Fry the onion in the melted butter until it is very soft. Stir in the curry powder and flour and cook for a few minutes. Add the apple and cook until tender. Add the rest of the ingredients and simmer for 20 minutes. Cool the sauce slightly, then pour into the goblet and blend for 30 seconds.
Reheat before serving.

Cucumber Sauce

Using Liquidiser or Blender

1 cucumber (cut up roughly)
½-ounce butter
½-ounce flour
¼ pint milk
2 tablespoons cream
2 tablespoons orange juice
Salt and pepper

Place all the ingredients in the goblet in the above order. Blend for 45-60 seconds, or until the sauce is smooth.
Transfer to a saucepan, bring to the boil and boil for 2 minutes, stirring all the time. Adjust seasoning before serving with either steamed plaice or salmon.

Mint Sauce

Using Liquidiser or Blender

1 small bunch of mint
2 level teaspoons sugar
1 level tablespoon boiling water } in a small basin
3-4 tablespoons vinegar

Remove the leaves from their stalks and place them in the goblet. Add the cooled sugar and water mixture and then the vinegar. Blend until the required texture is achieved (30-45 seconds approximately).

Apple Sauce

Using Liquidiser or Blender

1-pound cooking apples
1-ounce butter

Peel, core and slice the apples. Place in a small saucepan with a well-fitting lid. Add sufficient water just to cover the base of the saucepan. Cook on a low light until the apples are tender.
If there is a lot of liquid present, then strain this off before the apples are put into the goblet with the butter. Blend for approximately 30 seconds until the purée is formed.
Chill before serving with pork.

Tartare Sauce

Using Liquidiser or Blender

1 quantity of Mayonnaise (made with
   tarragon vinegar)
1 rounded tablespoon capers
3 gherkins
1 parsley sprig
1 hard-boiled egg (quartered)
2 tablespoons cream

Make the mayonnaise (page 46), then add the remaining ingredients. Blend for 10-15 seconds, and serve.

Bechamel Sauce

Using Liquidiser or Blender

Coating Consistency

$\frac{1}{2}$ pint milk
Bayleaf
Blade of mace
4 peppercorns
Piece of celery   } in a saucepan
Small piece of onion
Small carrot
1-ounce butter
1-ounce flour
Salt and pepper

Heat the contents of the saucepan to boiling point, remove from the heat and leave to infuse for 10 minutes.
Strain into the goblet, add the butter, flour and seasoning. Blend for 20 seconds, return to saucepan, then bring to the boil, stirring continuously. Boil for 2 minutes.

Pouring Consistency
Make in exactly the same way, but using $\frac{1}{2}$-ounce butter and $\frac{1}{2}$-ounce flour.

Syrup Sauce

Using Liquidiser or Blender

$\frac{1}{2}$-ounce cornflour or arrowroot
$\frac{1}{2}$-ounce sugar
1 level tablespoon golden syrup
$\frac{1}{2}$ pint juice (drained from fruit and made
   up with water if necessary)
1 teaspoon lemon juice

Place all the ingredients in the goblet and blend until smooth (15 seconds). Transfer to a saucepan, bring to the boil and then boil for 2 minutes, stirring all the time. Serve immediately.
This is delicious served with Fruit Upside-Down Pudding (page 38).

Barbecue Sauce

Using Liquidiser or Blender

1 small onion (sliced)
1-ounce butter
2 teaspoons flour
2 teaspoons french mustard
$\frac{1}{2}$ level teaspoon dry mustard
1 tablespoon Worcestershire sauce
1 teaspoon Tabasco sauce
$\frac{1}{2}$-ounce brown sugar
1 teaspoon salt
1 tablespoon vinegar
$\frac{1}{2}$ pint tomato juice

Cook the onion in the butter until soft and golden. Stir in the flour and cook for 2 minutes.
Place in the goblet with the rest of the ingredients and blend for 1 minute. Place the contents of the goblet in a saucepan, bring to boiling point and cook for 2 minutes stirring continuously. Keep hot until required.
Serve with grills, sausages or rissoles.

Mayonnaise

Using Liquidiser or Blender

1 egg
$\frac{1}{2}$ level teaspoon mustard
2 teaspoons vinegar
Salt and pepper
$\frac{1}{2}$ pint oil (corn oil or olive oil)

Break the egg into the goblet, add mustard, vinegar and seasoning. Run liquidiser at a medium speed and steadily pour the oil through the centre of the lid.

Using a Hand Mixer
Mix the egg, mustard, vinegar and seasoning. Then whisking on maximum speed, slowly add the oil to the mixture.

Above,
Curry Sauce, recipe page 44, and Mayonnaise

Caramel Sauce

Using Liquidiser or Blender

½ pint milk (heated to simmering point)
1-ounce butter
1½ level tablespoons cornflour
2 level tablespoons brown sugar
6 caramel toffees

Place all ingredients in the goblet and blend for 45 seconds. Pour the ingredients into a saucepan, bring to the boil and boil for 2 minutes, stirring all the time.
Serve with hot sponge puddings.

Orange Honey Butter

Using Liquidiser or Blender

6 strips of orange rind
3-ounce butter (softened)
6-ounce honey

Pare off thin strips of orange rind using a potato peeler. Place the ingredients in the goblet, blend until the orange rind is cut up and the butter and honey thoroughly mixed.
This is just as delicious if lemon is substituted for orange.
Serve with either baked or steamed sponge.

Sabayon Sauce

Using Hand Mixer

1-ounce castor sugar ⎫
2 egg yolks ⎪
2½ fluid ounce sherry ⎬ in a basin
2½ fluid ounce cream ⎭

Whisk all ingredients together over a saucepan of hot water. The sauce will thicken and become frothy.
Ideal for serving with Christmas Pudding.

Strawberry Whip

Using Liquidiser or Blender

6-ounce frozen strawberries (thawed)
¼-pound unsalted butter (softened)
2-ounce icing sugar (sieved)

Important
Be sure that the strawberries are thawed and the butter at least at room temperature.
Place all ingredients in the goblet in the above order. Blend until smooth. If necessary, stop blender during processing and push ingredients back towards the blades. Chill.
This makes a delicious sauce to be served with ice-cream.

Above, Strawberry Whip
Orange Honey Butter and Sabayon Sauce

French Dressing

Using Liquidiser or Blender

4 tablespoons oil
2 tablespoons wine vinegar
½ level teaspoon dry mustard
½ level teaspoon sugar
Salt and pepper

Place all ingredients in the goblet and blend for 30 seconds.

Basic White Sauce

Using Liquidiser or Blender

Coating consistency

1-ounce butter
1-ounce plain flour
½ pint milk

Pouring consistency

½-ounce butter
½-ounce flour
½ pint milk

Place all the ingredients in the goblet and blend for 30 seconds. Transfer the contents of the goblet to a saucepan, bring to the boil, then boil for 2 minutes, stirring continuously.

Variations
Sweet Sauce

1-2-ounce sugar
3 tablespoons rum (optional)
Add at the blending stage.

Mornay Sauce

2-ounce well-flavoured cheese (cubed)
Seasoning
Blend with other ingredients for 30-45 seconds.

Parsley Sauce

Few sprigs parsley
Seasoning
Add at the blending stage.

Onion Sauce

1 large onion (roughly chopped)
Seasoning
Add at the blending stage.

Cold Chocolate Sauce

Using Liquidiser or Blender

½ pint water   ⎫
6-ounce castor sugar ⎬ in a saucepan
2 egg yolks
4-ounce plain chocolate (broken into squares)
1 level teaspoon coffee
Rum to taste (optional)

Dissolve the sugar in the water and bring slowly to the boil. Boil vigorously for 2 minutes, then allow to cool slightly.
Place the egg yolks, chocolate and coffee in the goblet, pour in the syrup and blend for 15-20 seconds.
Return to the saucepan and heat gently to cook the egg yolks without boiling.
Chill and add rum if used. This is delicious with ice-cream.

Hot Chocolate Sauce

Using Liquidiser or Blender

½ pint milk
1 level tablespoon cornflour
2-ounce plain chocolate
1 level teaspoon instant coffee
2-ounce castor sugar
2 drops vanilla essence

Put all the ingredients except the vanilla essence in the liquidiser and blend for 30 seconds. Transfer to a saucepan and bring to the boil, stirring continuously. Boil gently for 2 minutes. Add the vanilla essence and serve.

Should you introduce your family to the delights of preparing drinks with the food mixer, you take the risk of being ousted from your kitchen! Few children can resist the challenge of preparing for themselves delicious Milk and Fruit Shakes and even fewer husbands will be able to resist trying out Orange Blossom, Winter Warmer and Orange Wine Cup. Let them go ahead; they can't go wrong with these simple recipes—and they might even let you taste the result.

Drinks

D

Milk Shake

Using Liquidiser or Blender

1 pint cold milk
1 individual size block of ice-cream
Sugar to taste

Flavour suggestions

2 teaspoons vanilla or any other essence
   plus colouring
1 level tablespoon instant coffee
1 level tablespoon cocoa
4-ounce fresh soft or frozen fruit (thawed)

Put all ingredients in the goblet and blend until smooth, approximately 30 seconds. Adjust sweetness at this stage if necessary.
Serves 2.

Raspberry Delight

Using Liquidiser or Blender

4-ounce raspberries (fresh or frozen)
1 pint cold milk
1-2 tablespoons sugar
Ice cubes (4-6)

Place all ingredients in the goblet and blend on maximum speed for 30 seconds. Strain and serve.
Serves 3.

Yoghurt Drink

Using Liquidiser or Blender

10 fluid ounce cold milk
1 carton flavoured yoghurt
1-ounce castor sugar

Blend all ingredients for 10 seconds.
Serves 2.

Whisky Cool

Using Liquidiser or Blender

2 fluid ounce whisky
$\frac{1}{2}$ an orange
2 rounded teaspoons castor sugar
3 ice cubes
Mint

Place the whisky and orange in the goblet and blend for 5 seconds. Strain. Blend the strained juice with sugar and ice cubes until the drink is frothy. Pour into a glass and decorate with mint.
Serves 1.

Strawberry Ice-Cream Soda

Using Liquidiser or Blender

4-ounce strawberries (frozen, fresh or tinned)
1 individual block vanilla ice-cream
1 tablespoon sugar
$\frac{1}{2}$ pint soda water
Pink colouring

Blend all ingredients in the liquidiser for 30 seconds. Strain and serve immediately.
Serves 2.

Blackcurrant Whip

Using Liquidiser or Blender

4 tablespoons blackcurrant purée
1 pint milk (cold)
Sugar to taste
Individual block of ice-cream (cut into cubes)

Place all ingredients except the ice-cream into the goblet and blend for 30 seconds. Chill thoroughly, then serve topped with ice-cream.
Serves 2.

Foaming Kenwood Lemonade

Using Liquidiser or Blender

1 lemon (thin skinned)
2 tablespoons sugar
6 ice cubes
$1\frac{1}{4}$ pints cold water
1 egg (complete with shell)

Place all ingredients in the goblet and blend for 7-10 seconds. Strain into a jug. To serve, add more ice cubes and float thin slices of lemon in the drink.
Serves 4.

Banana Egg Nog

Using Liquidiser or Blender

1 egg
1 banana
$\frac{1}{4}$ pint cold milk
2-ounce ice-cream
1 level tablespoon sugar

Blend all ingredients for 10 seconds and serve.
Serves 1.

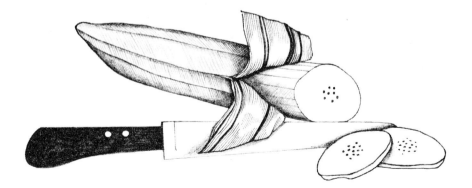

Orange Wine Cup

Using Liquidiser or Blender

2 medium oranges
1 pint medium white wine
2 fluid ounce brandy
2 level tablespoons castor sugar
1 pint soda water

Decoration
1 orange (thinly sliced)

Pare off the skin from the oranges very thinly using a potato peeler. Place this in the goblet with the wine, brandy and sugar. Blend for approximately 5 seconds. Strain the liquid off and chill this. Just prior to serving, add the juice of the oranges and the soda water.
Decorate with slices of orange.
Serves 6.

Golden Fizz

Using Liquidiser or Blender

1 tablespoon lemon juice
2 teaspoons castor sugar
1 egg
3 fluid ounce rum
Soda water

Place the lemon juice, sugar, egg and rum in the goblet and blend for 10 seconds. Divide between two glasses and top with soda water.
Serves 2.

Left, Orange Wine Cup and Golden Fizz

Opposite Left, Winter Warmer

Opposite Right, Fruit Shake, Iced Coffee and Choco-Peppermint Drink

Winter Warmer

Using Liquidiser or Blender

½ pint hot milk
1 egg yolk
1 fluid ounce sherry
½ fluid ounce brandy
Shake of nutmeg

Blend ingredients for 10 seconds. Serve immediately.
Serves 2.

Orange Blossom

Using Liquidiser or Blender

2 fluid ounce gin
Juice of 1 orange
2 teaspoons sugar
1 egg white
4 ice cubes

Place all ingredients in the goblet and blend for 5-10 seconds. Serve immediately in cocktail glasses.
Serves 2.

Fruit Shake

Using Liquidiser or Blender

8 fluid ounce milk
6 apricot halves (or an equal quantity of
    any tinned or soft fresh fruit)
½ small block ice-cream (cubed)

Blend fruit and milk for 30 seconds. Pour into a glass and decorate with cubes of ice-cream.
Serves 1.

Choco-Peppermint Drink

Using Liquidiser or Blender

2-ounce bar chocolate peppermint cream
    (broken into squares)
2 tablespoons boiling water
1 pint cold milk

Place the chocolate squares together with the boiling water in the goblet. Add the cold milk and blend for 15 seconds.
Serve immediately.
Serves 3.

Iced Coffee

Using Liquidiser or Blender

1 level tablespoon instant coffee
$\frac{1}{4}$ pint hot water
1 level dessertspoon sugar
6 ice cubes
Double cream

Place all ingredients except the double cream in the goblet in the above order. Blend until the mixture is smooth. Refrigerate for approximately 15 minutes. Pour into glasses, then add a tablespoon of cream to each, letting it form a layer on top.
Serves 2.

Hot Chocolate

Using Liquidiser or Blender

$\frac{1}{2}$ pint milk } in a saucepan together
$\frac{1}{2}$ pint water
2 level tablespoons drinking chocolate
1 teaspoon instant coffee
Sugar (optional)

Heat the milk and water together in a saucepan until they nearly reach boiling point. Cool slightly. Place the chocolate and coffee in the goblet, then pour on the milk and water. Blend for 10 seconds. This can be returned to the saucepan for re-heating if desired. Taste and add sugar before serving.
Serves 3.

Chocolate Ginger Cooler

Using Liquidiser or Blender

2 heaped teaspoons drinking chocolate
1 tablespoon boiling water
1 glass cold milk
1-2 tablespoons ginger wine

Decoration
Grated chocolate

Mix the drinking chocolate with the boiling water. Blend with the milk for 10 seconds. Add the ginger wine. Serve immediately topped with decoration.
Serves 1.

Chocolate Orange

Using Liquidiser or Blender

2 rounded teaspoons drinking chocolate } made into a paste and cooled
2 tablespoons boiling water
8 fluid ounce cold milk
Juice of 1 orange
2-ounce ice-cream

Place cool chocolate paste in the goblet together with the milk, orange juice and ice-cream. Blend on maximum speed for 10 seconds.
Serves 2.

The use of a table mixer takes all the hard work out of bread-making. The prospect of kneading batches of dough for long periods usually deters most housewives from yeast cookery, but the combination of Mixer and Dough-hook brings it back within her scope. If it has been a long time since you have made any bread, begin with Quick Wholemeal Bread. The result should soon restore your confidence. There are many interesting sweet and savoury doughs in this section that you can progress to and enjoy making. Don't forget that large batches of bread made with the Mixer are ideal for freezing.

Bread

Five Variations of White Bread

White Bread

*Using Table Mixer, Dough Hook
and Liquidiser*

1½-pound strong plain flour ⎫ sieved together
1 rounded teaspoon salt ⎭
½-ounce fresh yeast
¾ pint warm water (or milk and water mixed)
½-ounce soft lard

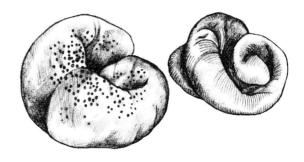

Place the bowl containing flour and salt in a warm place. Put the rest of the ingredients in the liquidiser goblet and blend together for 15 seconds.

Add the contents of the goblet to the flour and knead on speed 2 for 3 minutes or until the dough is smooth and leaves the sides of the bowl. Cover with a damp cloth and leave in a warm place until the dough has doubled in bulk. Knead again on speed 2 for 1 minute. Place the dough in two 1-pound loaf tins (well greased), or shape as required. Leave to prove in a warm place until the dough has risen to the top of the tin and is spongy to touch. Brush with milk and cook at 425°F/Reg 7 for 30-40 minutes. When cooked the loaf should sound hollow if tapped on the bottom. Cool on a wire rack.

Two Cottage Loaves

Divide the mixture into two halves for the two loaves. Take one half and cut it into two, one piece a little larger than the other, and make each into a round. Place the smaller piece on top of the larger piece and push the floured handle of a wooden spoon through the centre of both, and then remove it. Brush with egg or milk. Leave to prove 10-15 minutes and then bake at 450°F/Reg 8 for 25-30 minutes.

Two Poppy Seed Plaits

Divide the dough into two and cut one of the pieces into three. Roll each piece into a long, thin roll and place side by side. Gather the three ends together and form into a plait, tucking the ends underneath to neaten.

Repeat with the other piece of dough. Place plaits on a lightly greased and floured baking sheet, cover, and prove for 10-15 minutes. Brush with eggwash and sprinkle with poppy seeds. Place in the centre of the oven at 450°F/Reg 8 and bake for 30-35 minutes.

Crown Loaf

Complete dough sufficient to make three Crown Loaves.

Using 12-ounce of the dough, divide into six 2-ounce pieces. Roll each piece into a smooth ball. Place five in a greased 6 in. sandwich tin to form a ring, and the sixth in the centre. Brush with egg-wash and sprinkle with poppy seeds. If a crisp crust is required, brush with salt and water. Cover and leave to prove for 10-15 minutes. Bake just above the centre of a hot oven, 450°F/Reg 8 for 20-30 minutes.

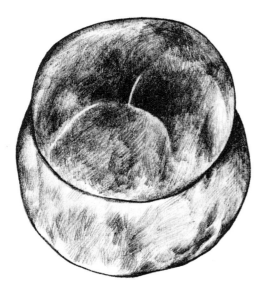

Cheese Loaf

Using Dough Hook and Slicer-Shredder

Makes 2 small loaves

½ pint warm water
½-ounce fresh yeast or 2 level teaspoons
  dried yeast
1 level teaspoon sugar
1-pound strong flour
2 level teaspoons salt
½ level teaspoon mustard
½ level teaspoon pepper
4-6-ounce finely grated cheese (on Slicer-Shredder)

Crumble the fresh yeast into the warm water and sugar and stir quickly. Place flour, salt, seasoning and three-quarters of the cheese in a bowl and pour the yeast liquid over the top. Switch on to minimum speed to incorporate the ingredients, thus forming the dough (approximately 1 minute). Increase the speed to 2 and knead for 2 minutes, until the dough leaves the sides of the bowl. Place dough in large, lightly greased polythene bag loosely tied, and leave until the dough has doubled in size and springs back when lightly pressed with a floured finger.

Reknead for 2 minutes on minimum speed. Cut the dough in half and shape (in the same way as with plain bread) to fit two greased 1-pound loaf tins. Place in a lightly greased polythene bag and prove dough until it reaches top of tin and springs back when lightly pressed with a floured finger. Sprinkle the remaining cheese over the tops of the loaves. Bake them on middle shelf in the oven on 400°F/Reg 6 for 45 minutes. Care should be taken not to overbake.

Variations

Cheese and Celery Loaf

Sprinkle an extra ounce of grated cheese mixed with 1 tablespoon celery salt on top of loaves before baking.

French Onion Bread

Bread with a delicious onion flavour for eating with salad or hot soup.

12 fluid ounce water
1-ounce fresh yeast or 1 level tablespoon
  dried yeast
1 level teaspoon sugar
½ level teaspoon salt
1-ounce lard
1½-ounce dried onion soup mix
1-pound strong white flour

Pour liquid into bowl. Crumble in fresh yeast and mix together on minimum speed for a few seconds. Add all other ingredients. Proceed as for Cheese Loaf.

Soda Bread

Using Table Mixer and Beater

1-pound wholemeal flour
½-pound strong plain flour
Pinch of salt } sieved in bowl
2 level teaspoons bicarbonate of soda
6 level teaspoons sugar
¾ pint milk

Place all dry ingredients in the bowl, add the milk and mix on minimum speed until it is incorporated. Increase to speed 3 and knead for 3 minutes to form a smooth dough that leaves the sides of the bowl clean.

Divide the dough into 2, shape each half into a round loaf. Place on a lightly greased baking sheet and bake at 400°F/Reg 6 for 50-60 minutes.

Quick Wholemeal Bread

Using Table Mixer and Dough Hook

Flowerpot loaves

1-ounce fresh yeast or 1 tablespoon dried yeast
¾ pint warm water
¾-pound wholemeal flour
¾-pound strong white flour
3 level teaspoons salt
3 level teaspoons sugar

The method of making White Bread may be used or this quicker method below:

Crumble the fresh yeast into the warm water and stir for a few seconds, OR sprinkle the dried yeast onto the warm water and leave until frothy (about 10-20 minutes) OR blend the dried yeast together with the warm water and sugar in the liquidiser for 5-6 seconds.

Place the flours, salt and sugar in the Chef bowl and pour the yeast/water mixture on top. Mix on speed 1 to incorporate the ingredients, thus forming the dough (approximately 1 minute). The dough should be soft but manageable. (It is often necessary to adjust the amount of flour slightly.) Knead for a further 2 minutes on speed 2 until the dough is smooth and elastic and the sides of the bowl are clean. The dough is now ready to use.

Above, Quick Wholemeal Bread

Opposite,
Orange Yeast Buns, Chelsea Buns and Doughnuts

Divide it into three and shape each piece to half fill a well-greased 5 in. earthenware pot, or three 1-pound loaf tins. Put to rise inside a greased polythene bag loosely tied, for 45-60 minutes, or until the dough has doubled in size and springs back when lightly pressed with a floured finger. Brush the tops with salt and water and sprinkle with lightly crushed cornflakes or cracked wheat. Bake on the middle shelf of oven on 450°F/Reg 8 for 30-40 minutes.

To prepare new flowerpots

Grease flowerpots well and bake empty in hot oven. Repeat several times. This treatment will prevent the loaves sticking, and for future use it is only necessary to grease pots in the usual way, as for tins.

For soft bread rolls which can be pulled apart

Prepare the 18 rolls and place ¾ in. apart on a baking tray—dust with flour. Bake near the top of the oven on 450°F/Reg 8 for 20-25 minutes.

Spicy Tea-Time Loaf

Using Table or Hand Mixer

6-ounce self raising flour ⎫ sieved together
1½ level teaspoons ground ginger ⎭
3-ounce butter or margarine
2-ounce walnuts (roughly chopped)
2-ounce dates (roughly chopped)
3-ounce soft brown sugar
1 egg ⎫ mixed together
4 tablespoons milk ⎭

Topping

1 rounded tablespoon soft brown sugar

Place the sieved flour and ginger in the bowl. Add the butter and mix on a medium speed until the mixture resembles breadcrumbs. Fold the nuts, dates and sugar in on the same speed, and when they are evenly distributed, add the liquid to form a stiff dough. Place the mixture in a 1-pound greased loaf tin. Sprinkle the tablespoon of soft brown sugar over the surface. Bake in a pre-heated oven on 350°F/Reg 4 for 50-60 minutes. Remove and cool on a wire tray.

Rich Yeast Mixture

Using Dough Hook

½-ounce fresh yeast ⎫ in bowl
4 fluid ounce warm milk ⎭
4-ounce plain flour (sieved)

Add the sieved flour to the milk and yeast mixture. Mix on speed 2. Knead for 2-3 minutes. Leave to double in size.

6-ounce plain flour (sieved)
½ level teaspoon salt
2-ounce castor sugar
1-ounce butter
1 egg (beaten)

Add the above ingredients to the risen dough and knead on speed 2 for a further 3 minutes. The dough should be well mixed at this stage and smooth. Use in a variety of ways—see following recipes.

Chelsea Buns

Using Dough Hook

Rich Yeast Mixture (this page)

Filling

1½-ounce brown sugar ⎫
1½-ounce butter or margarine ⎪
1 level teaspoon mixed spice ⎬ mixed together
½-ounce candied peel ⎪
1-ounce currants ⎭

Glaze (see Hot Cross Buns, page 65)

Roll the well-kneaded Rich Yeast Mixture into an oblong approximately 13 in. × 18 in.
Spread the filling over the whole of the dough. Roll up, starting at the long end—this should at this stage resemble a Swiss roll. Cut into 1½ in. slices using a sharp knife. Place in a deep-sided, greased tin so that they nearly touch. Put in a warm place to prove, 10-15 minutes. They should feel spongy to the touch at this stage.
Cook in a hot oven 400°F/Reg 6 for 15 minutes until brown. On removing from the oven, glaze immediately with sugar syrup.
Makes 12 buns.

Doughnuts

Using Dough Hook

½-pound Rich Yeast Mixture (page **63**)
2 tablespoons raspberry jam
Deep fat or oil for frying

Coating

1½-ounce castor sugar
½ level teaspoon cinnamon
½ level teaspoon nutmeg

Divide the Rich Yeast Mixture into eight pieces, and press each one flat. Place a little jam in the centre of each and bring the edges together to form a bundle, then roll to form a ball. Place on a lightly floured surface in a warm atmosphere until spongy to touch.

Heat the cooking oil until it begins to form bubbles round the handle of a wooden spoon. Place the doughnuts in the fat and fry till golden brown, turning frequently. Drain thoroughly on crumpled greaseproof paper. Place the sugar, cinnamon and nutmeg in a plastic bag and shake each doughnut in the mixture.

Makes 8 doughnuts.

Orange Yeast Buns

Using Dough Hook

½-pound Rich Yeast Mixture (page **63**)
Grated rind of an orange

Icing

6-ounce icing sugar (sieved)
6 teaspoons fresh orange juice
Few drops of orange colouring

Add the grated orange rind to the Rich Yeast Mixture and knead in well. Divide the mixture into eight and shape each one into a ball. Place on a lightly greased tray and leave in a warm place until they are spongy to touch (10-15 minutes). Bake at 400°F/Reg 6 for 15-20 minutes until lightly browned.

Whilst the buns are cooking, prepare a thick orange icing. Dip the warm buns in the icing, then place on a cooling tray—the icing will find its own level.

These are just as delicious if a lemon is used!
Makes 8 buns.

Hot Cross Buns

Using Table Mixer, Dough Hook and Beater

1 egg
Warm milk and water
1-ounce yeast
3-ounce butter or margarine (softened)
3-ounce castor sugar
1-pound plain flour (approx.)
$\frac{1}{4}$ teaspoon salt
1 level teaspoon mixed spice
1 level teaspoon cinnamon
$\frac{1}{2}$ level teaspoon ground cloves
4-ounce currants
2-ounce chopped peel

Cross Paste

1-ounce margarine (softened)
2-ounce plain flour
4 tablespoons water

Glaze

2-ounce sugar
4 tablespoons water

Break the egg into a measuring jug. Make it up to $\frac{1}{2}$ pint with warm milk and water. Dissolve the yeast in this liquid, and pour into the bowl. Add butter, sugar, flour, salt and spices in that order. Mix on minimum speed for a few seconds, increase to speed 1 and knead for 3 minutes until dough is smooth and elastic and the bowl is clean. (A little extra flour may be added at this stage to obtain the correct consistency.)

Remove the bowl, cover with a folded tea-towel, and leave to rise in a warm place until the dough doubles in bulk.

Add the fruit and reknead for 3 minutes on speed 1. Divide the dough into 16 pieces and shape into buns on a well-floured board. Place the buns on a greased tray well apart. Cover and leave in a warm place until the buns have doubled in size.

To make the crosses, place the margarine, flour and water in the Chef bowl and mix together using the K beater on speed 2. Pipe the crosses onto the buns having first brushed them with a little milk. Place in the centre of a hot oven (425°F/Reg 7) and bake for 15-20 minutes.

E

When the buns are cooked but still hot, place the glaze ingredients in a saucepan and dissolve over a low heat. Boil for about 2 minutes and brush on whilst hot.
Makes 16 buns.

Country Wholemeal Bread

Using Dough Hook and Liquidiser

1-pound stone-ground wholemeal flour ⎫ sieved into
1 rounded teaspoon salt ⎭ bowl
12 fluid ounce warm water (or water and
  milk mixed)
$\frac{1}{2}$-ounce fresh yeast
1 tablespoon extract of malt (optional)
Crushed corn (crushed in liquidiser)

Place the flour and salt in a warmed bowl. Put the warm liquid, yeast and malt extract in the liquidiser goblet and blend for 15 seconds. Pour the contents of the goblet into the flour and mix for 3 minutes on speed 2. The dough at this stage should be smooth and leave the sides of the bowl clean.

Shape either into a large bun or place in a 2-pound well-greased loaf tin. Cover and leave in a warm place until it doubles in bulk. Brush with milk and sprinkle liberally with crushed corn. Cook in a hot oven, 400°F/Reg 6, for approximately 30-40 minutes. The loaf should sound hollow when tapped on the bottom.

Apple Doughnuts

Using Table Mixer, Liquidiser and Dough Hook

1-pound strong flour
Pinch of salt
2-ounce castor sugar
2-ounce butter or margarine
1-ounce yeast (crumbled)
1½-pounds tart cooking apples
1 tablespoon lemon juice
1 standard egg

Sieve the flour, salt, and sugar into the bowl and add the butter, and crumbled yeast. Peel, core and quarter the apples. Drop the apple through the hole in the lid of the liquidiser goblet onto the revolving blades, adding some of the lemon juice at the same time. Continue feeding the apple through until the liquidiser is approximately one-third full. Empty the apple into the mixing bowl with the other ingredients and repeat the process with the apple until it is all dealt with, using the remaining lemon juice to prevent the apple from browning.

Add the egg and using the dough hook, knead on speed 1 until the ingredients are blended. Increase to speed 2 and then leave to rise until doubled in size, covered with a damp cloth.

Reknead the risen dough on speed 2 for 1 minute and turn onto a floured surface, moulding it into the shape of a ball. Divide into 16 pieces. Roll each piece into a ball using the palm of the hand. This is best done on an unfloured surface but with a floured hand, as long as the dough is not too sticky. Place the doughnuts on a lightly greased baking tray and set aside in a warm place until risen and puffy (about 25-30 minutes).

Deep fry in hot fat, turning them so that they brown evenly.

Drain, and roll in mixed castor sugar and cinnamon.

Makes 16 doughnuts.

Malt Loaf

Using Table Mixer with Dough Hook and Liquidiser

½-pound wholemeal flour ⎫
½-pound strong plain flour ⎬ sieved together
1 level teaspoon salt ⎭
½ pint warm water ⎫
½-ounce yeast │
½ level tablespoon soft brown sugar ⎬ in liquidiser
1 tablespoon malt extract │ goblet
½-ounce soft butter ⎭
3-ounce sultanas (optional)

Sieve the flours and salt into the bowl. Place water, yeast, sugar, malt and butter in the goblet and blend for 30 seconds.

Pour the liquid into the flour and knead on minimum speed until the flour is incorporated. Increase to speed 3 and knead for 3 minutes until the dough is smooth and leaves the sides of the bowl clean. Cover the bowl and leave in a warm place until the dough has doubled in size. Add the sultanas and reknead for 1 minute on speed 3.

Shape into two rounds and place on greased baking sheets. Prove until spongy to touch. Bake in a hot oven 425°F/Reg 7 for 35-40 minutes. The bread when cooked should sound hollow when tapped.

Lardy Cake

Using Table Mixer and Dough Hook

1-pound strong plain flour ⎫
2 level teaspoons salt ⎬ sieved together

½-ounce yeast ⎫
½ pint warm liquid (milk and water mixed) ⎬ mixed together
1 teaspoon sugar ⎭

Filling

4-ounce lard
4-ounce brown sugar
3-ounce currants

Syrup

1-ounce castor sugar ⎫ boiled together for 3 minutes
2 tablespoons water ⎭

Place the flour and salt in the bowl and add the yeast liquid. Knead on speed 3 for 3 minutes until the dough is smooth and leaves the sides of the bowl. Cover and leave in a warm place until the dough doubles in bulk.

Reknead on speed 3 for 1 minute and then roll out on a lightly floured surface until an oblong has been achieved (12 in. × 8 in. approximately). Mark the oblong into thirds. Place half the lard in flakes over the top two-thirds and then sprinkle on this half the currants and sugar. Fold into three, bringing the bottom third up first to form layers of dough and filling. Press the edges well with a rolling pin to seal them, give the dough a half turn and repeat with the rest of the filling. Roll the dough out to fit a 9 in. square greased cake tin. Score the top diagonally. Leave to prove in a warm place until it is spongy to touch (15-20 minutes).

Cook at 400°F/Reg 6 for 40-45 minutes. On removing from the oven, baste the top of the cake with any syrup in the tin. Whilst still warm, pour over the sugar syrup.
Cool before serving.

Savarin

Using Table Mixer and Beater

4-ounce strong plain flour ⎫
¼ level teaspoon salt ⎬ sieved together

4 fluid ounce milk (warm) ⎫
¼-ounce fresh yeast ⎬ together in small basin
1 level teaspoon sugar ⎭
1½-ounce butter or margarine (softened)
1 egg

Syrup

5 fluid ounce water
6-ounce castor sugar
2 tablespoons rum or liqueur

Apricot Glaze

3 tablespoons apricot jam ⎱ boiled together for
2 tablespoons water ⎰ 3 minutes

Decoration

Split almonds
¼ pint cream (whipped)

Place the sieved flour and salt in a bowl. Make a well in the centre and pour in the yeast liquid, butter and egg. Beat for 2 minutes on speed 3 using the beater. Pour into a greased 7 in. ring tin and leave (covered) to double in bulk. Cook for 15 minutes at 450°F/Reg 8, then turn down to 375°F/Reg 5 for a further 5-10 minutes until firm. Whilst the Savarin is in the oven, prepare the syrup by dissolving the sugar in the water over a low heat and boil for a few minutes. Remove from heat, allow to cool and then stir in the rum. Turn the Savarin on to a plate and whilst still warm spoon the syrup over until all is absorbed.

Brush all over with apricot glaze, stick the split almonds in the Savarin lengthways to achieve a porcupine effect, and decorate with whipped cream.
Serves 6.

Pizza

*Using Table Mixer, Dough Hook
and Liquidiser*

8-ounce wholemeal flour } sieved together in
1 level teaspoon salt } Chef bowl
$\frac{1}{4}$-ounce fresh yeast
$\frac{1}{4}$-pint warm water } in measuring jug
$\frac{1}{2}$-ounce soft lard
Salad oil for brushing

Pour the yeast mixture into the bowl containing flour and salt. Knead on speed 2 for 3 minutes, then cover with a damp cloth and leave in a warm place until the dough doubles in bulk.
Reknead for 2 minutes on speed 2.
Turn onto a floured board and roll into a long strip. Brush with salad oil and roll up like a Swiss roll. Repeat this twice, then divide into four. Roll each piece into a round approximately $\frac{1}{4}$ in. thick and cover with selected filling, seasoning well with salt, pepper and herbs. Brush generously with oil.
Bake on a heavy flat baking sheet for 15-20 minutes in a hot oven 450°F/Reg 8.
Serves 4.

Pizza Filling

For the fillings use a selection of the following. Quantities are for one pizza.

1 sliced tomato
1 tablespoon tomato purée
1-ounce cheese (grated)
1-ounce mushrooms (sliced)
2-ounce cooked ham
1 sausage (cut in slices)
Sardines
1-ounce onion (sliced)
Mixed herbs

Garnishes
Anchovy fillets
Strips of green pepper
Green olives
Stuffed olives
Gherkins
Bacon strips

Above, Pizza

Right, Bread Sticks

Bread Sticks

Using Table Mixer and Dough Hook

12-ounce strong plain flour } sieved together
1 level teaspoon salt
$\frac{1}{2}$-ounce yeast
7 fluid ounce warm milk } mixed together
$\frac{1}{2}$-ounce butter

Placed sieved flour and salt in bowl, add the yeast mixture to it and knead on speed 2 for 3 minutes, stopping to push the mixture back into the centre of the bowl if necessary.
Cover the bowl and leave in a warm place until the dough doubles in bulk. Reknead the dough on speed 1 for 2 minutes. Roll pieces of the risen dough into even lengths the thickness of a pencil. Place on lightly greased tin. Leave in a warm place to prove until spongy to the touch. Bake at 400°F/Reg 6 for 15-20 minutes until they are crisp and will snap in half.
Makes approximately 24 sticks.

A food mixer helps you make the lightest, melt-in-the-mouth scones and pastries because there is so little handling involved. One-Stage Pastry is a must for any housewife, for it can be prepared in minutes and then turned into delicious pies, pastries and flans. Use it as the basis for the Bakewell Tart and Savoury Courgette Flan next baking day—you will certainly be the toast of your family. Guests, too, will sing your praises when you serve home-made Profiteroles and Chocolate Eclair. Only you will know that the machine in the kitchen deserves much of the credit!

Scones and Pastry

Lattice Meat Tart, Choux Pastry, Tuna Fish Plait

Shortcrust Pastry

Using Table or Hand Mixer

8-ounce plain flour   ⎫ sieved together
1 level teaspoon salt   ⎭
4-ounce fat (a mixture of margarine and lard can be used)
8 teaspoons water

Place the sieved flour and salt in the bowl and add the fat, which should be cut up into small pieces. Commence on minimum speed until the ingredients are incorporated and then increase to speed 3 until the mixture resembles breadcrumbs. Switch off, sprinkle the water on top of the mixture and incorporate this on speed 3. The machine should be switched off as soon as the mixture comes together.

Use as required for pies, pasties, etc. This pastry is normally cooked at 425°F/Reg 7.

Almond Tarts

Using Table or Hand Mixer

4-ounce Shortcrust Pastry (this page)
1 tablespoon jam
1 egg white
1-ounce castor sugar
1-ounce ground almonds
2 drops of almond essence

Prepare the Shortcrust Pastry by following the instructions given above and using the egg yolk and water to bind it. Roll the pastry out thinly and cut rounds using a fluted cutter. Line 12 patty tins. Place a little jam in the base of each case. Whisk the egg white using a high speed until it is stiff and stands in peaks. Reduce the speed to minimum and fold in the sugar, ground almonds and almond essence. Place a teaspoon of the almond mixture in each tart. Bake at 350°F/Reg 4 for 25-30 minutes.
Makes 12 tarts.

Toffee Tart

Using Table or Hand Mixer

6-ounce Shortcrust Pastry (this page)

Filling

6-ounce margarine
6-ounce soft brown sugar
3½ fluid ounce milk   ⎫ blended together
1½-ounce plain flour   ⎭

Topping

2-ounce walnuts (chopped in liquidiser)

Roll the pastry out on a lightly floured board and line a 7 in. square tin. Bake the pastry case blind at 400°F/Reg 6 for 20-25 minutes.

Place the ingredients for the filling in a saucepan, bring to the boil and boil for 2-3 minutes stirring all the time. Pour this into the cooked pastry case and sprinkle liberally with nuts. Leave in a cool place to set.

This is delicious served with lightly whipped cream.
Serves 6-8.

Cherry Meringue Flan

Using Liquidiser

1 7-inch flan case baked blind
  (6-ounce Shortcrust Pastry, page 74)
1 14-ounce tin Cherry Pie filling

Confectioner's Custard

$\frac{1}{2}$ pint milk
4 level teaspoons cornflour
2 egg yolks
2 drops vanilla essence
2-ounce castor sugar

Meringues

2 egg whites
4 rounded tablespoons castor sugar

Place the ingredients for the Confectioner's Custard in the goblet and blend for 30 seconds. Transfer this to a saucepan, and bring to the boil. Allow to boil for 2 minutes, stirring all the time. Leave the mixture to cool and then pour into the prepared flan case. Allow the custard to set, then spread the cherry pie filling over.

Whisk the egg whites until stiff, then whisk in half the sugar—fold in the other half. With a large star tube, pipe a lattice pattern across the flan. Place under a hot grill to colour the meringue.
Serves 6.

Lattice Meat Tart

Using Table or Hand Mixer

8-ounce Shortcrust Pastry (page 74)

Filling

1-ounce margarine
1 onion ⎫
$\frac{1}{2}$ green pepper ⎬ chopped in liquidiser
1-ounce flour
$\frac{1}{4}$ pint stock
1 teaspoon Marmite or Bovril
1 tablespoon tomato purée
$\frac{1}{2}$-pound cooked meat (minced)
Salt and pepper

Garnish

Tomato
Parsley

Roll out the pastry to line a 7 in. square baking tin. Prick the base well and keep any scraps of pastry to make the lattice work decorations.

Melt the fat in a saucepan and gently fry the onion and pepper until tender. Add the flour and cook for a few minutes, then stir in the stock, Marmite and tomato purée to make the sauce. Bring to the boil, add the meat and then adjust the seasoning before pouring into the pastry case. Roll out the rest of the pastry and cut into strips. Twist these and put across the top of the tart to form a lattice pattern. Bake at 350°F/Reg 4 for 35-40 minutes until pastry is golden brown. Serve either hot or cold.
Serves 4.

Savoury Courgette Flan

Using Slicer-Shredder and Liquidiser

Shortcrust Pastry

6-ounce plain flour
1½-ounce margarine
1½-ounce lard
Salt and pepper to taste
6 teaspoons cold water

Filling

½-pound courgettes (topped, tailed, and sliced)
½-pound tomatoes
1-ounce margarine
1 onion (skinned and chopped roughly)
2 eggs
3 fluid ounce milk
Salt and pepper to taste
2-ounce cheese (grated)

Line a 7 in. flan ring with the Shortcrust Pastry (page 74). Prick the base all over and bake blind for 10 minutes at 425°F/Reg 7.

Blanch the tomatoes and cut each tomato into half-a-dozen pieces. Place the margarine, courgettes, tomatoes and onion in a saucepan with a lid over a low heat and steam fry for 10-15 minutes shaking the saucepan frequently.

Place the eggs, milk, salt and pepper in the goblet and blend for 5 seconds.

Fill the baked flan case with the steam fried courgettes, tomatoes and onions. Pour over the eggs, milk, salt and pepper. Sprinkle the grated cheese over the top and cook in a moderate oven 375°F/Reg 5 for 20 minutes or until firm.
Serves 6.

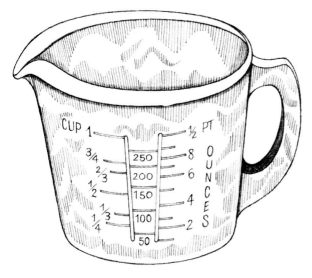

Bakewell Tart

Using Table or Hand Mixer

Pastry

4-ounce plain flour ⎫ sieved together
Pinch of salt ⎭
½-ounce castor sugar
2-ounce butter or margarine
4 teaspoons cold water

Filling

2 rounded tablespoons jam
4-ounce ground almonds
2-ounce ground rice or semolina
4-ounce castor sugar
2 eggs
2-ounce butter or margarine (softened)
¼ teaspoon almond essence

Prepare pastry following the method for Short-crust Pastry (page 74). Roll out on a lightly floured board until the pastry is large enough to fit a 7 in. deep-sided flan ring or sandwich tin. Line the ring or tin with the pastry and trim the edges. Spread jam on the bottom of pastry. Place the rest of the ingredients in the bowl and combine on a low speed until all are thoroughly mixed. Place on top of jam and spread evenly. Bake at 325°F/Reg 3 on the middle shelf for 1 hour 10 minutes approximately. Serve either hot or cold.
Serves 6-8.

One-Stage Pastry

Using Table or Hand Mixer

5-ounce margarine (softened) or 2½-ounce
  lard and 2½-ounce margarine
2 tablespoons cold water
8-ounce plain flour (sieved)
Salt to taste

Place the margarine, cold water, 4-ounce of the flour, and the salt in the bowl. Incorporate the ingredients on speed 1 and cream them for approximately 15 seconds until well mixed.
Add the remaining flour and mix on speed 1 for a further 15-20 seconds or until it has become a firm dough. Knead the dough and roll it out, then use as required.
This pastry is usually cooked at 425°F/Reg 7.

Sweet Pastry Flan Case

Using Table or Hand Mixer

4-ounce plain flour
2-ounce butter (cut up roughly)
½-ounce castor sugar
1 egg yolk ⎫ mixed together
4 teaspoons water ⎭

Place the flour in a bowl, add the cut-up butter and mix on a low speed until the mixture resembles breadcrumbs. Add the sugar and then the egg and water; continue to mix on the low speed but switch off as soon as all ingredients are incorporated. Cover the pastry and allow to relax in the refrigerator for 15 minutes.
Roll the pastry out on a lightly floured board to form a round. Line a 7 in. fluted flan ring with the pastry. Prick the bottom well. Place a piece of foil in the bottom and then bake blind at 400°F/Reg 6 for 20-25 minutes. The flan ring can be removed approximately half way through the cooking to give the outside a chance to brown.

Rough Puff Pastry

Using Table Mixer

2½-ounce margarine
2½-ounce lard
½-pound plain flour } sieved together
1 level teaspoon salt
5 fluid ounce water } mixed together
1 teaspoon lemon juice

Mix the margarine and lard together with a fork. If they become very soft, then place in the refrigerator to become firm.

Place the sieved flour and salt in the bowl, add the fat which should be in pieces the size of brazil nuts. Mix on minimum speed just long enough to coat the fat in flour. Add the water and lemon juice. Continue to mix on the same speed until the water is incorporated. Turn onto a well-floured board and roll into an oblong 12 in. × 9 in. approximately. Mark the oblong into three, fold the bottom over the middle section and then the top section down. Give the pastry a half turn and repeat the rolling and folding three more times. It is possible to roll the pastry out more thinly each time—it should be ¼ in. thick on the last rolling. If possible, leave the pastry to relax in the refrigerator between rollings. After the last rolling, leave the pastry for at least 30 minutes in the refrigerator before using for sausage rolls, meat or fish pies, or Tuna Fish Plait (this page).

Rough Puff Pastry should be cooked at 400°F/ Reg 6.

Tuna Fish Plait

Using Table Mixer

Filling

7-ounce tin of tuna fish (drained)
2 teaspoons tomato purée
2 teaspoons anchovy essence
1 level teaspoon dried or fresh thyme
1 shake of tabasco
1 clove of garlic } chopped together
½ level teaspoon salt
1-ounce butter

½ recipe for Rough Puff Pastry (this page)

Place all ingredients for the filling in the bowl and mix on a medium speed until smooth.

Roll the pastry on a lightly floured board to an oblong 12 in. × 8 in. approximately. Mark into three lengthways and make diagonal slashes ½ in. apart down the two sides. Fold the strips of pastry across the filling, using them alternately. Brush with milk and bake at 400°F/Reg 6 for 30-35 minutes or until golden brown.

Serve either hot or cold.

Serves 4.

Fish Envelope

Using Liquidiser or Blender

4-ounce Rough Puff Pastry (this page)
½-pound cooked white fish
¼ pint coating sauce (page 48)
Seasoning
1 hard-boiled egg (quartered)

Remove skin from fish, flake this and add to sauce. Season well and allow to cool.

Roll the pastry to a 9 in. square on a lightly floured board. Place on a baking tray. Place the cooled fish mixture plus egg into the centre. Damp the edges of the pastry and fold into an envelope shape, sealing edges well. Any trimmings can be used to make leaf decorations. Brush with egg and bake in a hot oven (450°F/Reg 8) for 15 minutes and then for a further 30 minutes at 400°F/Reg 6.

Serves 4.

Choux Pastry

Using Table or Hand Mixer

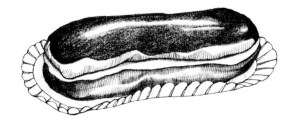

¼ pint water ⎤ in a saucepan
2-ounce margarine ⎦
2½-ounce plain flour (sieved onto paper)
2 standard eggs

Bring the water and fat slowly to the boil. Remove from heat as soon as it has boiled and beat in the sieved flour. Cool for a few minutes.

If using a table mixer, transfer the mixture to the bowl and beat in the eggs a little at a time until the paste is soft and just holds its shape.

The hand mixer can be used to add the eggs to the mixture in the saucepan.

Use in any of the following ways:

Choux Puffs

Using Table or Hand Mixer

1 quantity of Choux Pastry (this page)
¼ pint cream (whipped)
1-ounce icing sugar (sieved)

Spoon small rounds of Choux Pastry onto a deep, lightly greased baking tray (a meat tin is ideal). Cover with another tray giving them each room to rise. Bake in a pre-heated oven on 400°F/Reg 6 for 20-25 minutes or until firm. Place on a wire rack and split immediately. When completely cold, fill with cream and serve sprinkled with icing sugar.

In summer these make a delicious sweet with fresh strawberries mixed with the cream.
Makes approximately 12 puffs.

Chocolate Eclairs

Using Table or Hand Mixer

1 quantity of Choux Pastry (this page)
¼ pint cream (whipped)
Chocolate glacé icing

Using a large plain piping tube, pipe the Choux Pastry into 3 in. lengths on a lightly greased baking tray. Cook in a pre-heated oven on 400°F/Reg 6 for 20-25 minutes until golden brown.

Split with a sharp knife or pair of scissors, then put on a wire rack to cool. When thoroughly cold, fill with whipped cream and coat the top with chocolate glacé icing.
Makes 10 eclairs.

Profiteroles

Using Table or Hand Mixer

1 quantity of Choux Pastry (this page)
¼ pint cream (whipped)
Cold Chocolate Sauce (page 48)

Place teaspoons of the Choux Pastry onto lightly greased baking sheets. Bake in a pre-heated oven on 400°F/Reg 6 for 15-20 minutes. Cool on a wire tray.

When cool, split each one and fill with cream. Arrange in a pyramid on a large serving dish or in individual sundae dishes. Pour over the chocolate sauce and serve.
Serves 6.

Scones

Using Table or Hand Mixer

8-ounce self raising flour (sieved)
Pinch of salt
2-ounce butter or margarine (cut into pieces)
1-ounce castor sugar
1 egg
4 tablespoons milk (approx.)

Place flour and salt in the bowl, add the fat and mix on minimum speed until the fat is broken up. Increase to speed 3 until the fat is evenly distributed. Add the sugar and mix in thoroughly. Reduce the speed to minimum, adding the egg, and then the milk a little at a time until a soft, pliable dough is formed.
Turn onto a lightly floured board and knead until smooth. Roll out to ½ in. thickness and then cut out with a 2 in. cutter. Transfer to a lightly greased baking tray. Brush the tops with milk.
Bake towards the top of a hot oven (425°F/Reg 7) for approximately 10 minutes.

Variations

Fruit—Add 2-ounce of the fruit of your choice together with the sugar.

Spice Scones—Add 1 level teaspoon of cinnamon together with the sieved flour.

Lemon and Orange Scones—Add the grated rind of lemon and orange with the sugar.

Makes approximately 12 scones.

Horseshoe Scone

Using Table or Hand Mixer

Use basic Scone recipe (this page) plus:
2-ounce dried fruit
1-ounce soft brown sugar
1-ounce butter or margarine (softened)
1 level teaspoon cinnamon

Roll the scone mixture out on a lightly floured board to a rectangle 8 in. × 12 in. approx. Spread with butter and then sprinkle fruit, sugar and cinnamon onto this. Beginning at the long end, roll this up to form a sausage.
Transfer to a lightly greased baking sheet and bring the ends towards one another to form a horseshoe shape. Slash the top with a knife, brush with milk, and bake at 375°F/Reg 5 for 20-25 minutes.
Allow to cool on the baking tray for a few minutes and then transfer to a rack so that it can become completely cold before serving.
Serves 6.

Left, Various Scones

Above Right, Horseshoe Scone

F

Swedish Meat Ring

Using Table or Hand Mixer

8-ounce self raising flour ⎱
Pinch of salt ⎰ sieved together
3-ounce margarine
1 egg (made up to ¼ pint with milk)

Filling

1 medium onion (chopped)
1-ounce margarine
1-ounce flour
¼ pint stock or gravy
8-ounce minced beef (cooked)
Seasoning

Sieve the flour and salt into the bowl. Rub in the margarine using speed 3 until it resembles bread-crumbs. Add enough liquid to form a soft scone dough (any egg and milk which is left can be used for glazing).

Fry the onion gently in the margarine until it is tender and golden brown. Add the flour and cook for a few minutes. Add the stock, bring to the boil ensuring that the sauce is free from lumps, and then add the minced beef. Season well and allow to cool.

Roll the dough out on a lightly floured board to an oblong 8 in. × 12 in. Spread the cool filling over evenly. Roll up starting at the long end to form a ring. Place on a lightly greased baking tray. Cut the ring at 2½ in. intervals nearly to the centre, then twist the slices so that they overlap and the filling shows. Brush with eggs and milk. Bake at 400°F/Reg 6 for approximately 30 minutes. Serve hot or cold.

Serves 4.

Cornish Pasties

Using Table or Hand Mixer

Pastry

½-pound shortcrust pastry

Filling

½-pound chuck steak
4-ounce potato
4-ounce carrot
2-ounce onion
Seasoning

Prepare the Shortcrust Pastry by following the instructions on page 74. Roll the pastry out and cut out 2 circles the size of dinner plates, or 4 the size of tea plates.

Cut both the meat and vegetables into small cubes and season. Divide the filling between the pastry rounds. Place the filling in the centre of rounds. Brush the pastry edges with water. Bring edges up together and over filling and press gently together all round to seal. Twist the edges neatly with the fingertips. Place on an ungreased baking sheet and brush with milk. Bake at 400°F/Reg 6 for the first 20 minutes then reduce to 350°F/Reg 4 for a further 30 minutes.

Serve either hot or cold.

The only problem in producing biscuits and cookies so quickly with your food mixer is that your family will soon discover how delicious they are and demand more. When you realise, however, that one minute's mixing can produce two dozen Brown Sugar Crunchies, you'll soon see that there's no excuse for not keeping those tins filled. All the recipes in this section are so quick and easy that you can double up on quantities and store the finished result in air-tight containers. Making large amounts becomes routine when you have a Mixer to help.

Biscuits and Cookies

Viennese Fingers and Cookies

Viennese Fingers

Using Table or Hand Mixer

4-ounce butter or margarine
1-ounce icing sugar (sieved)
4-ounce plain flour (sieved)
$\frac{1}{2}$ teaspoon vanilla essence

Filling

2-ounce icing sugar (sieved) ⎫
1-ounce butter or margarine ⎬ in a warmed bowl
1 drop vanilla essence ⎭

Chocolate Coating

3-ounce chocolate (melted over hot water)

Place the butter and icing sugar in a warmed bowl and cream until the mixture is quite soft. Add the flour and vanilla essence and continue to beat until the mixture is fairly soft but smooth.
Transfer the mixture to a large piping bag fitted with a star nozzle. Pipe 3 in. fingers onto a lightly greased baking sheet. Cook at 350°F/Reg 4 for 10-15 minutes or until they just begin to change colour. Cool on a wire rack.
Beat the filling ingredients together until they are smooth and creamy; the chocolate can be put to melt at the same time. When cool, sandwich the fingers together with the cream and then dip the ends of the biscuits in chocolate. Leave to dry completely before storing.
Makes 6 fingers.

Date and Walnut Bars

Using Table Mixer and Liquidiser

8-ounce plain flour (sieved)
4-ounce butter or margarine (softened)
4-ounce castor sugar
1 egg yolk
3 tablespoons raspberry jam ⎫
1 teaspoon lemon juice ⎬ mixed together
4-ounce dates
2-ounce almonds (blanched)
2 egg whites
2-ounce castor sugar

Place the flour, butter, sugar and egg yolk in a bowl and mix on speed 1, gradually increasing to speed 3 until the ingredients are thoroughly mixed. Pack into the bottom of a greased 8 in. square tin and prick with a fork. Bake at 350°F/Reg 4 for 20-25 minutes or until golden brown.
Remove from oven and spread the raspberry jam and lemon juice over the top. Using the liquidiser, chop the dates and almonds by dropping them through the hole in the lid onto the revolving blades. Beat the egg whites until foamy and gradually add 2-ounce castor sugar. Continue to beat until the foam is stiff. Fold the chopped almonds and dates into the egg whites and spread over the jam. Bake at 350°F/Reg 4 for a further 25 minutes. Cool in tin and then cut into bars.
Makes 16 bars.

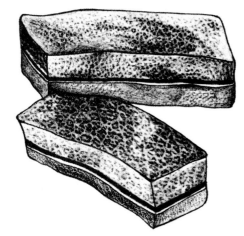

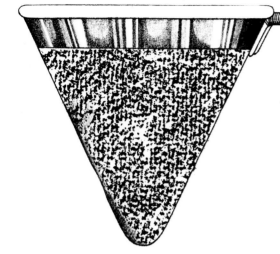

Chocolate Peanut Squares

Using Table or Hand Mixer

6-ounce plain flour (sieved)
4-ounce butter or margarine (softened)
2-ounce granulated sugar
2-ounce self raising flour ⎱
Pinch of salt ⎰ sieved together
2-ounce roasted peanuts (chopped in liquidiser)
4-ounce Polkadots or chocolate chips
4-ounce soft brown sugar
$\frac{1}{2}$ teaspoon vanilla essence
2 eggs
Icing sugar (for dredging)

Place the plain flour, butter and sugar in the bowl and switch on to a low speed until the ingredients have come together (up to 1 minute). Press into the bottom of a 9 in. square tin and bake at 325°F/ Reg 3 for 25 minutes.

Place the self raising flour and salt in the bowl, add the rest of the ingredients and mix thoroughly. Spread evenly over the baked layer and then return to the oven at 325°F/Reg 3 for a further 35 minutes. Cool slightly and cut into squares and dredge with sieved icing sugar before serving.
Makes 16 squares.

Spicy Date Flapjacks

Using Liquidiser or Blender

3-ounce dates (stoned)
6-ounce sugar
1 teaspoon mixed spice
6-ounce margarine
6-ounce rolled oats
1-2 tablespoons syrup
1 teaspoon lemon juice (optional)

Drop the dates and then the sugar through the hole in the lid of the liquidiser onto the revolving blades and blend for 10 seconds. Place these together with the remaining ingredients in a saucepan over a low heat, and stir until the margarine has melted. Tip the ingredients into a small greased Swiss roll tin and cook at 350°F/ Reg 4 for 30 minutes or until golden brown. Loosen the flapjack from the edge of the tin and mark where it is to be cut.

Leave to cool in the tin for $\frac{1}{2}$ hour, then cut and store in an airtight tin.
Makes 18 fingers.

Florentines

Using Table Mixer and Liquidiser

3-ounce mixed nuts (walnuts, brazils,
  blanched almonds or hazelnuts)
3-ounce glacé cherries
½-ounce angelica
4-ounce self raising flour (sieved)
4-ounce demerara sugar ⎫
4-ounce butter or margarine ⎬ in saucepan
4-ounce golden syrup ⎭

Place the nuts, cherries and angelica in the liquidiser goblet, switch on and allow to run until they are roughly chopped. The motor may have to be stopped and the ingredients pushed back onto the blades. Place the chopped ingredients in the bowl together with the sieved flour. Melt the ingredients in the saucepan but do not allow to become too hot. Pour the contents of the saucepan onto the dry ingredients in the bowl and mix on a low speed until they are thoroughly incorporated. Place teaspoonsful of mixture well apart on baking sheets which have been lightly greased. Bake in a pre-heated oven on 325°F/Reg 3 for 10-15 minutes. Cool on a wire tray. If liked, coat the backs with melted chocolate.
Makes approximately 14.

Cheesy Tomato Snacks

Using Table or Hand Mixer

4-ounce plain flour ⎫
½ level teaspoon salt ⎬ sieved together
Pinch of mustard ⎭
2-ounce butter or margarine
2-ounce cheese (finely grated)
4 teaspoons cold water
3 medium tomatoes (cut into rings)
1-ounce cheese (finely grated)

Place the sieved ingredients in the bowl and add the butter. Mix on a low speed until the mixture resembles breadcrumbs. Add the cheese and then the water. Knead to a smooth dough on a lightly floured board, prick all over. Roll out thinly and cut out with a 2 in. pastry cutter. Top each with a slice of tomato and a little grated cheese.
Bake on a lightly greased baking sheet at 350°F/ Reg 4 for 10-15 minutes or until the biscuit is lightly coloured.
Note. Mixed herbs could be added to the cheese sprinkled on top to add to the flavour.

Oaties

Using Table or Hand Mixer

4-ounce plain flour ⎫ sieved together
1 level teaspoon bicarbonate of soda ⎭
4-ounce rolled oats
3-ounce golden syrup
3-ounce sugar
3-ounce margarine
3 tablespoons water

Place all the ingredients in the bowl (in the above order) and mix on speed 3 until they are all incorporated. Take heaped teaspoonsful of the mixture, shape into rounds and place on a lightly greased baking sheet. Cook at 325°F/Reg 3 for 15-20 minutes. Allow to cool completely on a rack before storing in an airtight tin.
Makes approximately 24 biscuits.

Iced Almond Biscuits

Using Table or Hand Mixer

7-ounce plain flour ⎫ sieved together
1-ounce cornflour ⎭
4-ounce castor sugar
4-ounce butter or margarine
2-ounce ground almonds
1 egg yolk ⎫
2 tablespoons water ⎬ beaten together
½ teaspoon almond essence ⎭

Coating

1 egg white
4-ounce icing sugar (sieved)
2-ounce almonds (chopped in liquidiser)

Place sieved flours, castor sugar and butter in the bowl and mix on speed 3 until the mixture resembles breadcrumbs. Add the ground almonds and then the liquid (do not pour all in at once). Continue to mix until a soft dough is formed. Extra liquid can be added at this stage if necessary. Knead lightly on a floured surface, then roll out to approximately ¼ in. in thickness. Cut out with a 2½ in. cutter.

Mix the egg white with the icing sugar on speed 3. Place a teaspoonful on each biscuit. Sprinkle on a few chopped nuts. Place on a lightly greased baking sheet and cook at 350°F/Reg 4 for 15-20 minutes or until just pale gold in colour.
Makes approximately 24 biscuits.

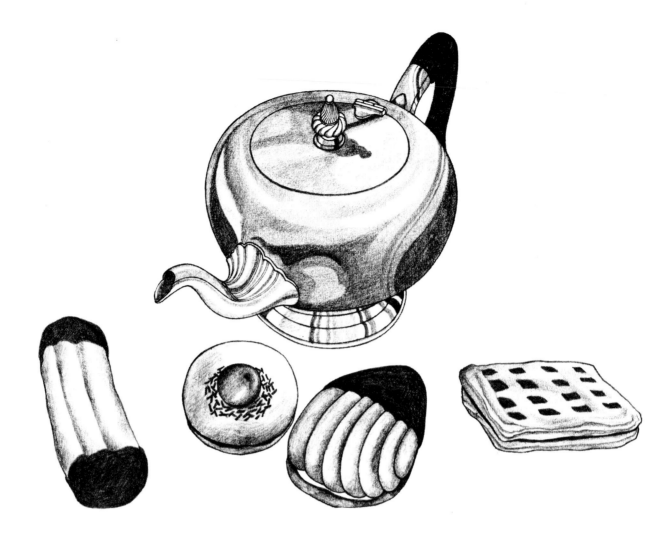

Chocolate Melting Moments

Using Table or Hand Mixer

3-ounce butter or margarine (softened)
3-ounce icing sugar (sieved)
2-ounce self raising flour (sieved)
1-ounce cocoa (sieved)
3-ounce rolled oats
4-ounce plain chocolate (melted)

Follow the method for the Melting Moments (page 91), adding the cocoa at the same time as the flour and rolled oats. When they are cooked, leave them to cool on the baking tray. Melt some plain cooking chocolate and spread it over the top of the cooled biscuits with a knife to leave a rough effect.
Makes 12 biscuits.

Nutty Cookies

Using Table or Hand Mixer

2-ounce margarine
2-ounce brown sugar
4-ounce plain flour
1 tablespoon clear honey
1-ounce nuts (chopped)
1-ounce chocolate chips

Cream the margarine and sugar until light and fluffy, then fold in the flour. Mix in the honey, nuts and chocolate.
Divide the mixture into 9, shape into rounds and place on a greased baking sheet.
Cook at 350°F/Reg 4 for 20-25 minutes until golden brown. Store in an airtight tin when cool.
Makes 9 cookies.

Melting Moments

Using Table or Hand Mixer

3-ounce margarine or butter (softened)
3-ounce icing sugar (sieved)
2-ounce self raising flour (sieved)
3-ounce rolled oats
Glacé cherries

Place the margarine and icing sugar in the bowl and cream them together starting with speed 1 and gradually increasing to speed 3 until the ingredients are combined. Add the flour and 1-ounce oats and mix on speed 1 until the ingredients have come together to form a soft pastry texture. Roll into a sausage and cut into 12 equal pieces. Roll each piece into a ball. Dip them into water and then into the remaining 2-ounce rolled oats. Place them onto a greased baking tray 3 in. apart and flatten them slightly. Bake in the centre of the oven on 350°F/Reg 4 for 15-20 minutes. Place glacé cherries in the centre of each biscuit and cool on the tin. Store in an airtight container.
Makes 12 biscuits.

Welsh Cakes

Using Table or Hand Mixer

8-ounce plain flour }
½ level teaspoon salt } sieved together
4-ounce margarine or butter
4-ounce castor sugar
8-ounce currants
1 egg (lightly beaten)
4 tablespoons milk (approx.)

Sieve the flour into the bowl, add the fat and mix on a low speed until the mixture resembles bread-crumbs. Using the same speed mix the sugar and currants in thoroughly. Add the egg and sufficient milk to form a firm dough. Roll out on a lightly floured surface to ½ in. thickness and cut out with a round cutter (2 or 3 in. in diameter). Cook on a medium hot griddle until brown on each side.
Serve hot sprinkled with castor sugar, or cold spread with butter.
Makes 12 cakes.

Tea-Time Bars

Using Table Mixer and Liquidiser

6-ounce plain flour (sieved)
4-ounce margarine
2-ounce sugar
2-ounce margarine (softened)
Juice of 1 lemon
Juice of 1 orange
½-pound stoned dates
4-ounce chocolate (melted)

Place the flour and 4-ounce margarine in the bowl and mix on speed 2 until the mixture resembles coarse breadcrumbs. Using the same speed, stir in the sugar. Press this mixture into the bottom of a well-greased small Swiss roll tin. Bake at 350°F/Reg 4 for 20 minutes.
Place the margarine (2-ounce) and juice of the fruits in the goblet and blend for approximately 15 seconds. Drop the dates onto the revolving blades through the hole in the lid. The mixture may have to be stopped occasionally to push the food back onto the blades. Spread the date mixture onto the cooked base and return to the oven at the same temperature for 15 minutes.
When completely cold cover the surface with melted chocolate. Cut into squares when set.
Makes 16 squares.

Chocolate Caramel Fudge

Using Table or Hand Mixer

4-ounce butter or margarine
2-ounce castor sugar
4-ounce self raising flour (sieved)

Caramel Filling

4-ounce margarine or butter ⎫
1 small tin condensed milk ⎬ in a saucepan
2 tablespoons golden syrup ⎪
4-ounce castor sugar   ⎭

Decoration

4-ounce plain chocolate (melted)

Cream margarine and sugar together on speed 3 until light and fluffy, reduce to minimum speed and incorporate the flour. Press the mixture into a lightly greased small Swiss roll tin. Bake at 350°F/Reg 4 for 20 minutes or until lightly browned.

Filling

Heat the filling ingredients over a low heat and bring slowly to the boil. Allow the mixture to boil for 4-5 minutes, stirring all the time. Cool a little and then pour over the cooked layer, spreading evenly. Allow this to become completely cold.
Cover the caramel filling with melted chocolate. When the chocolate is set, the fudge can be cut into fingers and served.
Makes 18 fingers.

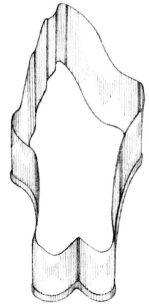

Brown Sugar Crunchies

Using Table or Hand Mixer

6-ounce soft brown sugar
6-ounce self raising flour
½ level teaspoon bicarbonate of soda
1 egg
2-ounce almonds (coarsely chopped)
4-ounce chocolate (chopped) or Polkadots
3-ounce butter or margarine (softened)
1 teaspoon vanilla essence

Place all the ingredients in the mixing bowl and mix on speed 3 until the ingredients are well incorporated. Put teaspoonsful of the mixture on a lightly greased baking sheet. Bake at 350°F/Reg 4 for approximately 10 minutes.
Allow to cool on the baking tray for a few minutes and then transfer to a cooling rack.
Store in an airtight tin.
Makes 20-24 biscuits.

Chequered Biscuits

Using Table or Hand Mixer

12-ounce plain flour (sieved)
4-ounce castor sugar
8-ounce soft butter or margarine
1 rounded tablespoon cocoa (sieved)

Place the flour, sugar, and butter in the warmed bowl and mix on speed 3 until a soft dough is formed. Divide the mixture into two, leaving one half in the bowl. Add cocoa to one half and mix on speed 3 until thoroughly mixed.
Divide both plain and chocolate mixture in two and then on a lightly floured surface roll each piece into a long sausage (approximately 16 in.). Place the four pieces together to form a chequered roll and then cut off fairly thick pieces to form biscuits.
Place on a lightly greased baking sheet. Cook at 350°F/Reg 4 for 15-20 minutes. Cool thoroughly before serving.
Makes 20-24 biscuits.

Almond Meringues

Using Table or Hand Mixer

2 egg whites
1-ounce ground almonds
1 level teaspoon almond essence
Fat for deep frying
Castor sugar (approx. 2-ounce)

Whisk the egg whites on maximum speed until they reach meringue peak. Reduce to speed 1, fold in the ground almonds and almond essence, then switch off immediately.
Heat the fat to 350°F or until a 1 in. cube of bread turns golden brown in 60 seconds. Drop one table-spoon of the almond mixture at a time into the fat and cook until golden brown. Drain well, toss in castor sugar and serve either hot or cold.
Makes 8-10 meringues.

Gone are the days when cake-making resulted in aching arms from long periods of beating the ingredients. Today you can let the machine do the work for you. It is not only quicker but the results are better too. Try Marmalade Fruit Cake and Quick Treacle Loaf to prove to yourself and your family that cakes without effort taste just as good.

Cakes

From Left to Right, Swiss Roll
Battenburg and Dr May's Fruit Cake

Fatless Sponge

Using Table Mixer

3 eggs
3-ounce castor sugar
3-ounce self raising flour (sieved)

Place the eggs and sugar in the warmed bowl and whisk on speed 6 until the mixture is very light and thick. (When ready, the mixture should be so thick that the whisk leaves a firm imprint.) Remove the whisk from the planet hub and use this to stir in the sieved flour.

Cook in two 6½ or 7 in. lightly greased, floured tins at 350°F/Reg 4 for 25-30 minutes. Cool on a rack.

For two 7½ or 8 in. tins, use 4 eggs, 4-ounce castor sugar and 4-ounce S.R. flour.

Swiss Roll

Grease and line a large Swiss roll tin. Make up a 3-egg quantity of sponge mixture and place immediately in the tin. Bake at 425°F/Reg 7 for 8-10 minutes. Turn immediately onto a sugared greaseproof paper, remove lining paper and trim edges, if they are crisp.

Spread with softened jam and roll immediately. If a cream Swiss roll is to be made, roll up immediately this comes from the oven. Unroll when cool, spread with cream and re-roll.

Battenburg

Using Table or Hand Mixer

4-ounce margarine or butter
4-ounce castor sugar
2 eggs
4-ounce self raising flour (sieved)
1 level tablespoon cocoa   } blended together and
1 tablespoon boiling water } cooled

Decoration

2 tablespoons jam (softened)
1 (8-ounce) packet made-up almond paste

Make up the cake mixture as for Victoria Sponge (see page 102). Grease and line two 1-pound loaf tins. Divide the mixture in half and add the cocoa blend to one lot. Place the mixture in the two tins and cook at 350°F/Reg 4 for 20-25 minutes. Allow each cake to cool in the tin for a few minutes, and then turn onto a wire rack.

Trim the edges of cake and cut each into two lengthways strips, making four in all. Join alternate colours together with jam to make a square. Roll the marzipan to an oblong approximately 7 in. × 11 in. Spread cake with jam, place on centre of marzipan. Press marzipan around the cake, sealing underneath, trim the ends and crimp the edges.

Make a 7 in. plait with marzipan trimmings and place on top.

Mocha Fudge Fingers

Using Liquidiser or Blender

½-pound Marie biscuits (crumbed in liquidiser)
2-ounce walnuts (chopped in liquidiser)
3-ounce dates (roughly chopped)
½ teaspoon vanilla essence ⎫
1 rounded teaspoon cocoa ⎪
1 rounded teaspoon instant coffee ⎬ in a basin
4-ounce plain chocolate (broken into ⎪
   squares) ⎪
1 small tin condensed milk ⎭

Place the biscuit crumbs, nuts and dates in the bowl and mix together thoroughly. Stand the basin containing the rest of the ingredients over a saucepan of hot water. Stir until the chocolate is melted.
Pour this mixture onto the dry ingredients and mix thoroughly. Turn the mixture into a lightly greased 8 in. square tin and press down smoothly. Chill in the refrigerator for several hours, then cut into fingers and serve.
Makes approximately 16 fingers.

Dr May's Fruit Cake

Using Table or Hand Mixer

1-pound self raising flour (sieved)
½-pound margarine or butter
½-pound castor sugar
12-ounce mixed fruit
1 egg and water to make up
   to 8 fluid ounce

Place the sieved flour in a bowl, add the margarine, and switch to a low speed. When the mixture resembles breadcrumbs, add the sugar and fruit and mix in thoroughly.
Add the egg and water. Place the cake mixture in an unlined 8 in. greased round cake tin. Bake at 350°F/Reg 4 for approximately 1½ hours.

Quick Treacle Loaf

Using Table Mixer

2-ounce shelled walnuts (chopped slightly)
10-ounce self raising flour ⎫
1 level teaspoon baking powder ⎬ sieved
¼ level teaspoon salt ⎭
4-ounce margarine (softened)
4-ounce castor sugar
4-ounce sultanas
2 standard eggs
4-ounce black treacle
¼ pint milk

Place all the ingredients in the bowl in the order listed above. Incorporate on speed 1 (10-15 seconds approximately), and then increase to speed 3 to make a total mixing time of 1 minute, or until all ingredients are well mixed. Turn into a greased 2-pound loaf tin. Bake at 325°F/Reg 3 for 1 hour or until firm and deep brown in colour. Allow to cool in the tin for 10 minutes. Turn out of tin and cool on a rack.
Note. This loaf improves by keeping in an airtight container for a few days.

Optional Topping

1 orange
3-ounce castor sugar

For a change, take the thin orange portion of the peel and the fruit of the orange together with the sugar, and drop this through the hole in the lid of the liquidiser onto the revolving blades. Blend for 10 seconds. Pour this mixture over the cake as soon as it comes from the oven. Cool the cake in the tin before removing.

Feathered Marble Cake

Using Table or Hand Mixer

4-ounce margarine or butter
4-ounce castor sugar
2 eggs
4-ounce self raising flour
Red and green colouring

Icing

6-ounce icing sugar (sieved)
Warm water to mix
Red and green colouring

Make up the sponge following instructions for Victoria Sponge (page 102). Divide the mixture into three, colouring one part green, another pink, and leaving the last part plain. Place the mixture in blobs in a prepared 8 in. round or square tin. Bang the tin lightly to ensure that the mixture is evenly distributed. *Do not spread with a knife.* Cook at 350°F/Reg 4 for 30-35 minutes. Cool thoroughly on a tray.

Decoration

Make up icing (it should thickly coat the back of the spoon when ready) and spread it evenly on the cake. With a fine paint brush, draw lines of red and green colouring alternately across the cake. Draw a skewer through the lines of colour to give a marbled or feathered effect. Leave the icing to set firmly before the cake is served.

Top, Feathered Marble Cake, All-In-One Meringue and No Bake Chocolate Cake

Above, Christmas Cake

No Bake Chocolate Cake

Using Liquidiser or Blender

8-ounce Digestive biscuits (crumbed)
2-ounce butter or margarine
2 rounded tablespoons golden syrup or honey
4-ounce Bournville chocolate (broken into squares)
2-ounce marshmallows (quartered)
1-ounce chopped nuts

Crumb the biscuits by dropping them through the hole in the lid of the liquidiser onto the revolving blades. Melt the butter, syrup and chocolate in a saucepan over a low heat. Stir in the biscuit crumbs and nuts, then finally the marshmallows. Turn into a lightly greased 7 in. sandwich tin or a greased 7 in. flan ring on a baking sheet. Press down firmly. Place in the refrigerator until completely set. Turn out of tin and cut into small portions before serving.
Makes 12 slices.

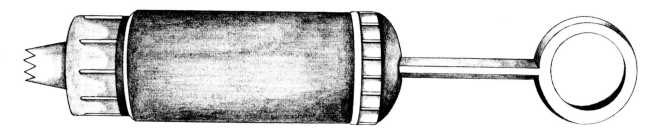

Christmas Cake

Using Table Mixer

10-ounce butter
10-ounce soft brown sugar
Grated rind of 1 lemon
6 eggs
1½ tablespoons black treacle
11-ounce plain flour
1½ level teaspoons mixed spice ⎱ sieved together
½ level teaspoon nutmeg ⎰
4-ounce glacé cherries
4-ounce mixed cut peel
1-pound currants
10-ounce sultanas
6-ounce raisins
4-ounce chopped nuts (almonds or walnuts)

Prepare a 10 in. round or 9 in. square cake tin by lining with a double thickness of greased paper. Warm bowl and beater.
Place butter and sugar and grated lemon rind in the bowl and switch to minimum speed until ingredients have combined, then increase to speed 2. Cream until light and fluffy, switch off, add treacle and turn to speed 2 again.
Add the eggs one at a time, thoroughly beating each. Reduce the speed to minimum and tip in all sieved flour, spices, then fruit and nuts. Switch off as soon as ingredients are incorporated and turn into the tin.
Bake in a slow oven (290°F/Reg 1) for approximately 4 hours. (Check after 3 hours.)
When cake is cold it can be turned upside down, pierced with a skewer and brandy poured in. The cake should then be stored in an airtight tin.

All-In-One Meringue

Using Table or Hand Mixer

4 egg whites
9-ounce icing sugar (sieved)

Place egg whites and icing sugar in the Mixer bowl and whisk on speed 1 for a few seconds until ingredients are incorporated.
Switch to maximum speed and continue to whisk until the meringue mixture is very thick (about 10 minutes). (A firm impression should be left with the whisk at this stage.)
Using a star nozzle, pipe the mixture onto a lightly greased lined tray into the shapes required. Cook at the lowest oven setting possible until the meringues are thoroughly dry. (This will take several hours, but the time depends on the size of the meringues.)
Once thoroughly dry and cool, the meringues can be stored in an airtight tin and then used as required.

Hand Mixer

If using a Hand Mixer, then the mixture should be whisked over a saucepan of hot water. The whisking time will then be reduced considerably.

Orange Frosted Chocolate Gâteau

Using Table Mixer

4-ounce margarine or butter (softened)
4-ounce castor sugar
3-ounce self raising flour ⎫
1-ounce cocoa ⎬ sieved together
1 level teaspoon baking powder ⎭
2 large eggs

Decoration

Butter icing—orange (page 110)
Orange Frosting (half of the recipe on page 110)
Chocolate curls

Place all the cake ingredients in the bowl in the order listed above. Incorporate the ingredients on speed 1 (approximately 10 seconds) and then increase to speed 4 to make a total mixing time of 1 minute. Turn into two 6½ or 7 in. sandwich tins which have been greased and lined. Bake at 350°F/Reg 4 for 25-30 minutes. Turn out of the tins immediately and cool on a rack.

Fill the centre of the cake with orange Butter Icing and frost the outside of the Gâteau with Orange Frosting and decorate with chocolate curls, made by cutting shavings off a block of chocolate with a potato peeler.

Victoria Sponge

Using Table or Hand Mixer

4-ounce butter or margarine
4-ounce castor sugar
2 eggs
4-ounce self raising flour (sieved)

Flavour Suggestions

Chocolate—Blend 1 rounded tablespoon of cocoa with 2 tablespoons of boiling water. Allow this paste to cool before using.
Coffee—1 tablespoon coffee essence or strong black coffee.
Orange—1 tablespoon undiluted orange squash or grated rind of 1 orange.
Lemon—1 tablespoon undiluted lemon squash or grated lemon rind.

Warm the bowl and beater thoroughly. Place the butter and sugar in the bowl and cream this on speed 3 until light and fluffy.

Add the eggs a little at a time, beating thoroughly after each addition.

Reduce to minimum speed and add the flour and any flavouring. Switch off immediately the flour has been incorporated.

Turn into a greased, lined, deep 8 or 8½ in. tin and bake at 350°F/Reg 4 for 30-35 minutes. Allow to cool in the tin for a few minutes and then cool completely on a rack.

Note. 6-ounce butter, sugar, flour and 3 eggs are required when two 7 or 7½ in. tins are being used.

Special Sponge

Using Table Mixer

6-ounce castor sugar
3 eggs (large)
6-ounce self raising flour (sieved twice)
2-ounce butter or margarine ⎤ melted together over
1 small tablespoon golden syrup ⎬ a low heat and
2 tablespoons water ⎦ left to cool

Warm the bowl and whisk. Place the castor sugar and then eggs in the bowl and whisk on maximum speed until it is pale and thick (approximately 5 minutes). Switch off and sift the flour over the surface. Fold the flour in lightly with a spatula or a metal spoon, and then carefully fold in the other ingredients which have been allowed to cool. Place the mixture in two greased, lined and floured 8 or 8½ in. sandwich tins. Bake in a 350°F/ Reg 4 oven for approximately 20 minutes or until the cakes are golden brown and firm to the touch. Remove from tins immediately and leave to cool on a wire tray.

Try filling this sponge with Soured Butter Icing (page 111) and decorate with Crystallised Caramel Chips (page 112).

One Stage Victoria Sandwich

Using Table or Hand Mixer

4-ounce margarine (soft)
4-ounce castor sugar
2 eggs
4-ounce self raising flour ⎤ sieved together
1 level teaspoon baking powder ⎦

Place all ingredients in the bowl in the above order. For the Chef, beat on speed 1 for 10 seconds and then speed 8 for 50 seconds.

If using a hand mixer, mix on speed 1 for 30 seconds and top speed for 1 minute 30 seconds. Transfer the mixture to either one 8 in. greased tin or two 7 in. tins.

Bake in a pre-heated oven at 325°F/Reg 3. The 8 in. cake will require 35-45 minutes; the two 7 in. tins 25-35 minutes. Cool on a wire tray.

Dundee Cake

Using Table or Hand Mixer

6-ounce margarine
6-ounce castor sugar
Rind of ½ an orange
4 eggs
7-ounce plain flour ⎫
½ level teaspoon baking powder ⎬ sieved together
1 level teaspoon mixed spice ⎭
6-ounce currants
6-ounce sultanas
6-ounce raisins
2-ounce glacé cherries (quartered)
3-ounce mixed peel

Decoration

2-ounce almonds (blanched)

Beat margarine, sugar and grated orange rind until pale and fluffy. Add eggs one at a time, beating well after the addition of each one. If the mixture curdles slightly then a tablespoon of flour can be added with the last egg. Fold in the flour and other dry ingredients on a low speed, then add the prepared fruit. Transfer the cake mixture to a greased and lined 7 in. round cake tin. Decorate with almonds and cook at 300°F/Reg 2 for 2-2½ hours.

Note. If a hand mixer is used, then this can be used as far as the addition of fruit, but this should be done by hand.

Farmhouse Fruit Cake

Using Table or Hand Mixer

10-ounce self raising flour ⎫
½ level teaspoon cinnamon ⎬ sieved together
6-ounce soft brown sugar
6-ounce butter or margarine (cut up into pieces)
8-ounce mixed dried fruit
2-ounce walnuts (chopped in liquidiser)
2-ounce glacé cherries (quartered)
2 eggs made up to ¼ pint with milk

Place flour, cinnamon and brown sugar in the bowl, add butter or margarine. Mix on speed 2 until the mixture resembles breadcrumbs. Reduce to minimum speed and add the rest of the dry ingredients. When the ingredients are thoroughly incorporated, add the egg and milk. Transfer the mixture to a greased lined 7½ in. round tin. Bake on the centre shelf of a moderately hot oven 350°F/Reg 4 for 1 hour 45 minutes. Allow to cool in the tin for a few minutes and then transfer to a cooling rack. Store in an airtight tin.

Marmalade Fruit Cake

Using Table Mixer

6-ounce margarine or butter (softened)
6-ounce castor sugar
8-ounce plain flour ⎫
½ level teaspoon baking powder ⎬ sieved together
½ level teaspoon cinnamon ⎭
3 large eggs
10-ounce mixed dried fruit
2 tablespoons chunky marmalade
½ level teaspoon finely grated orange peel

Place all ingredients in the bowl in the order listed above. Incorporate the ingredients on speed 1 (10-15 seconds approximately) and then increase to speed 3 to make a total mixing time of 1 minute, or until all ingredients are well mixed.

Turn into a greased and lined 8 or 8½ in. cake tin. Bake at 325°F/Reg 3 on the middle shelf for 1½-1¾ hours or until a skewer inserted into the centre of the cake comes out clean. Leave it in the tin for 2-3 minutes, then turn out and cool on a wire tray.

Date and Honey Loaf

Table Mixer with Liquidiser

8-ounce dates (stoned)
¼ pint milk
3 level tablespoons honey (clear)
1 tablespoon water
1 standard egg
8-ounce self raising flour ⎫
½ level teaspoon baking powder ⎬ sieved together into a mixing bowl
¼ level teaspoon nutmeg ⎭
Honey to glaze

Place dates, milk, honey, water and egg into the goblet and switch on to maximum speed for 10 seconds. Pour these ingredients into the bowl with the flour, baking powder and nutmeg and mix on speed 2 for 15 seconds. Pour into a greased 1-pound loaf tin and cook for approximately 1-1¼ hours at 350°F/Reg 4 until firm and browned.

Leave to cool in tin for 10 minutes, turn out onto a wire cooling rack and brush the top with honey to form a glaze. The flavour will improve by keeping for a week or two in an airtight tin.

Note. This loaf can be served sliced and buttered for a change.

Prune and Apricot Cake

Using Table or Hand Mixer

4-ounce dried apricots
4-ounce prunes (stoned)
8 fluid ounce milk or water
6-ounce castor sugar or soft brown sugar
3-ounce margarine
1 egg
½ level teaspoon cinnamon ⎫
¼ level teaspoon nutmeg ⎪
¼ level teaspoon salt ⎬ sieved into the bowl
8-ounce plain flour ⎪
1 level teaspoon baking powder ⎭

Place the apricots, prunes, milk, sugar, margarine and egg in the liquidiser goblet and switch on to maximum speed for 20 seconds or until the dried fruit is finely chopped. Pour this mixture onto the sieved ingredients and mix on speed 3 for 15 seconds or until the ingredients are well mixed. Pour this into a 2-pound loaf tin and bake in a 350°F/Reg 4 oven for 1 hour or until it is firm and golden brown, and a skewer inserted into the middle of the cake comes out clean. Cool completely on a wire rack. Store in an airtight tin.

Simnel Cake

Using Table Mixer

Marzipan

8-ounce ground almonds
4-ounce icing sugar (sieved)
4-ounce castor sugar
1 egg
½ teaspoon almond essence

Cake

6-ounce butter or margarine
6-ounce demerara sugar
3 eggs
8-ounce plain flour ⎫
1 level teaspoon nutmeg ⎪
1 level teaspoon cinnamon ⎬ sieved together
1 level teaspoon mixed spice ⎪
Pinch of salt ⎭
1 tablespoon milk or sherry
8-ounce currants
6-ounce sultanas
6-ounce raisins
6-ounce glacé cherries (quartered)
2-ounce mixed peel

To Decorate

⅔ of marzipan
1 egg yolk
Easter decorations (sugar eggs, chicken)
Apricot glaze

Glacé icing

4-ounce icing sugar (sieved)
Water to mix

Mix the ground almonds and sugar on a low speed. Add the egg, then the essence a little at a time, and mix to form a smooth, stiff paste.
Divide into three, roll out one piece to a 7½ in. circle to fit the tin, cover the rest to use later.
Warm the bowl and beater and cream the butter and sugar until light and fluffy. Add the eggs one at a time, beating thoroughly after each addition. Reduce to minimum speed, add sieved flour, spices, liquid, then fruit, switching off as soon as

it is incorporated. Place half the mixture in a greased and lined 7½ in. tin, cover with the circle of marzipan and then add the remaining mixture. Bake in a slow oven 300°F/Reg 2 for approximately 3½ hours. Allow to cool and store until ready to decorate.
Roll one of the remaining portions of marzipan to a 7½ in. circle.
Brush the top of the cake with apricot glaze (1 tablespoon jam and 1 teaspoon water boiled together for 2 minutes). Place the glazed surface of the cake onto the marzipan, press firmly and trim edges. Turn the cake the right way up. Divide the last portion of marzipan into 11 balls and place them around the edge of the cake. Brush with egg yolk and brown under the grill.
When thoroughly cooled, prepare glacé icing and pour onto centre of cake.
Decorate with 'Happy Easter', fluffy yellow chick or sugar eggs. Traditionally, the sides of the cake should be covered with yellow and white ribbon.

Choc Nut Loaf

Using Table or Hand Mixer

4-ounce soft brown sugar
4-ounce butter or margarine
3 eggs
8-ounce self raising flour (sieved)
2-ounce walnuts (chopped in liquidiser)
2-ounce chocolate (roughly chopped)

Decoration

1-ounce walnuts (very roughly chopped)

Cream the sugar and butter together until they are creamy and light in colour. Add the eggs a little at a time, beating well when each part is added. (If curdling occurs at this stage then a heaped teaspoon of the measured flour can be added.) Switch to minimum speed to fold in the flour, nuts and chocolate—switch off as soon as they are incorporated. Turn into a greased, lined 1-pound loaf tin and sprinkle nuts over the top. Bake at 350°F/Reg 4 for approximately 1¼-1½ hours.

The simplest Victoria Sponge can be made into an exciting gâteau at a moment's notice when those unexpected visitors arrive, simply by adding delicious icings and fillings. Coffee and Chocolate Butter Icings for filling and frosting take roughly three minutes to prepare, so you can serve a Mocha Cake within minutes. Let the food mixer be your kitchen assistant, and you'll never be caught with nothing to give your guests. Both Almond Paste and Royal Icing are made simple with the aid of a Mixer, and of course it's a great help at Christmas time when decorating the cake has to be fitted in with so many other jobs. If you want a new way of finishing off your celebration cakes, try the Moulded Icing by way of a change.

Icings and Fillings

Orange Glacé Icing, Chocolate Cream and Butter Icing

Butter Icing

Using Table or Hand Mixer

3-ounce butter or margarine (softened)
6-ounce icing sugar (sieved)
1 tablespoon cold water

Variations

Chocolate—1 rounded tablespoon cocoa plus 2 tablespoons boiling water. Allow the paste to cool before using.
Coffee—1 tablespoon coffee essence or strong black coffee.
Orange—1 tablespoon orange juice or undiluted squash.
Lemon—1 tablespoon lemon juice or undiluted squash.

Place the butter and water or flavouring in the bowl and blend on speed 3 until well mixed. Add the icing sugar, blend on minimum speed until incorporated, and then turn to speed 3 for a few minutes until the cream is light and fluffy.

Orange Glacé Icing

Using Table or Hand Mixer

Thinly pared-off peel of an orange
   (no pith)   } in a saucepan
3 tablespoons water
6-ounce icing sugar (sieved)
Few drops of orange colouring

Place the saucepan containing orange peel and water on a low heat, bring to the boil and then remove from heat. Leave to infuse for 10 minutes. Mix 2 tablespoons of the water into the icing sugar on minimum speed, gradually increasing the speed as the mixture thickens. The icing should thickly coat the back of a spoon. Add colouring if necessary, and then use immediately. The peel may be cut into thin strips and used for decoration.
Note. For quickness try 2 tablespoons of undiluted orange squash with the icing sugar.

Orange Frosting

Using Table or Hand Mixer

1-pound granulated sugar
Juice of 1 orange made up to $\frac{1}{4}$ pint with water
Orange colouring
Whites of 2 standard eggs
Pinch of salt

Place the sugar, juice and orange colouring in a saucepan over a low heat, and stir until the sugar has dissolved. Bring it to the boil. Allow to boil fairly briskly without stirring for approximately 5 minutes (or until a few drops of the mixture in a cup of cold water form a soft ball when gently rolled between finger and thumb). The temperature of the syrup at this stage should be 238°F if a sugar thermometer is being used. Meanwhile beat the egg whites and salt to a stiff foam. When the syrup has reached the soft ball stage, pour onto the whisked egg whites in a slow, steady stream with the mixer on maximum speed. Continue whisking until the frosting is cool enough and thick enough to spread. The frosting hardens very quickly when it has cooled, and should therefore be used immediately.
Note. This recipe provides enough frosting to cover the top and sides of three 7 in. sandwich cakes or one deep 7-8 in. cake cut into two layers.

Peppermint Chocolate Icing

Using Table or Hand Mixer

2-ounce plain chocolate (broken into squares)
2 tablespoons water
1-ounce butter
6-ounce icing sugar (sieved)
3 drops peppermint essence

Warm chocolate and water in a small saucepan over a low heat until dissolved. Pour onto butter and sugar and beat till smooth, then add peppermint essence to taste.
Sufficient to cover a 6 in. cake.

Chocolate Cream

Using Table or Hand Mixer

1 level tablespoon instant coffee ⎫ in a saucepan
3 tablespoons water ⎭
8-ounce plain chocolate (broken into squares)
2 egg yolks
¼ pint double cream (whipped)
Few drops of vanilla essence

Dissolve the coffee in the water over a low heat. Remove from the heat and add the chocolate squares, mixing until they dissolve. Beat in the egg yolks until the mixture is smooth and creamy. Leave to cool for 10 minutes. When cool, gently fold in the whipped cream and essence on a low speed.

Royal Icing

Using Table or Hand Mixer

2 egg whites
1-pound icing sugar (sieved)
1 teaspoon lemon juice
1 teaspoon glycerine

Place egg whites in the bowl and switch on to minimum speed. Gradually add icing sugar, then lemon juice and glycerine. Turn to medium speed until icing is smooth and white.
Cover with a damp cloth and leave to release air bubbles before using.
This icing is to be used for flat icing or piping.
Note. If a hand mixer is used, then this is the maximum amount to be prepared.

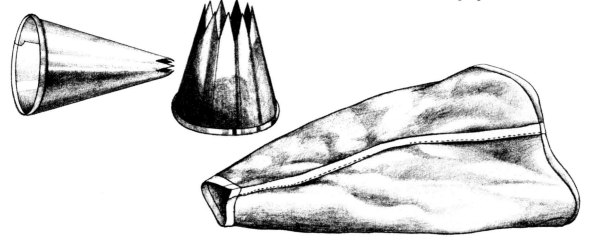

Soured Butter Icing

Using Table or Hand Mixer

4-ounce margarine or butter (softened)
8-ounce icing sugar (sieved)
⅛ pint soured cream

Place all ingredients in the bowl and bring together on a low speed. Increase the speed to beat the icing until it is light and fluffy.

Chocolate Tutti Frutti Filling

Using Table or Hand Mixer

1 quantity of Chocolate Cream (this page)
1-ounce walnuts (coarsely chopped in liquidiser)
1-ounce glacé cherries (quartered)

Mix the nuts and cherries into half the Chocolate Cream and use this to sandwich together two halves of a sponge cake. Use the other half of the cream to coat the outside of the cake to make a rich and delicious gateau.

Crystallised Caramel Chips

Simple but effective decoration

3 tablespoons sugar
1 tablespoon water

Place the ingredients in a saucepan over a low heat and stir until the sugar has turned to a toffee colour (approximately 4 minutes). Pour the caramelised sugar onto a piece of tin foil about 9 in. square and leave to set for a few minutes. Crack it and place pieces of the 'chips' at random on top of an iced cake.

St Clement's Icing

Using Table or Hand Mixer

Rind of ½ orange and lemon
6-ounce icing sugar (sieved)
1 tablespoon each of orange and lemon juice
1-ounce butter (softened)

Grate the orange and lemon rind onto the icing sugar in a bowl with the fruit juice and butter. Beat together until smooth and creamy.
This can be used for icing sponge sandwich cakes.

Almond Paste

Using Table Mixer

6-ounce icing sugar (sieved)
6-ounce castor sugar
12-ounce ground almonds
1 egg (lightly beaten)
1 dessertspoon lemon juice
1 dessertspoon brandy or rum

Top, Crystallised Caramel Chips and St Clements Icing

Above, Ingredients for Almond Paste

Place the sugars and ground almonds in the bowl and mix on minimum speed for a few seconds until they are thoroughly combined. Add the egg, lemon juice and spirit and mix on medium speed until a smooth paste is formed. Knead lightly on a sugared board before using as required.

Satin Frosting

Using Table or Hand Mixer

2 egg whites
14-ounce granulated sugar
Pinch of salt
4 tablespoons water
2 level teaspoons cream of tartar

Place all ingredients except the cream of tartar in a bowl over hot water, whisk for 5 minutes. Take off heat and cool for approximately 5 minutes. Whisk again until it is shiny and stands in peaks. Alternatively, place all ingredients except the cream of tartar in the warmed bowl of the Chef or Major and whisk on maximum speed for 15 minutes. When stiff, add the cream of tartar. There is sufficient to frost the outside of a 7 in. cake.

Meringue and Cream Icing

Using Table or Hand Mixer

2 egg whites
3-ounce icing sugar (sieved)
¼ pint double cream (whipped)

Optional: 1 or 2 teaspoons rum, sherry or
    brandy can be added by way of a special
    filling.

Place egg whites in bowl and whisk until foamy. Add a little icing sugar with the mixer still running and continue adding a little more icing sugar until it is all incorporated. Whisk until it forms stiff peaks and gradually add whipped cream. The rum, sherry or brandy can be added at this stage also.

H

Moulded Icing

Using Table Mixer

This is placed directly on the cake and then decoration applied to it.

1-pound icing sugar (sieved)
1 egg white
1 tablespoon liquid glucose

Place all ingredients in the bowl and bring together on a low speed. Increase to a medium speed and beat for approximately 10 minutes. Remove from the bowl and knead by hand until smooth. Roll out on a board sprinkled with icing sugar, then place on the cake and mould using one's fingers until it is perfectly smooth. Once the whole cake has been covered, the icing can be made to shine by gently rubbing with silk.
Left-over icing can be coloured and used for moulded decoration.
Sufficient to cover a 7 in. cake.

Chocolate Syrup Topping

Using Hand Mixer

2-ounce chocolate (plain or milk)
1-ounce butter or margarine
2 tablespoons golden syrup

Break the chocolate into squares and place in a bowl to melt over a saucepan of hot water. Add the butter and syrup and beat well until the mixture cools and a spreading consistency is reached.

Many of the best traditional recipes in this country have been passed down from generation to generation and, luckily for us, they have lost none of their excellence. Think, as you enjoy Potato Cakes for breakfast or Shortbread for tea, how much Grandma would have appreciated the labour-saving food mixer!

Also in this section, you will find some traditional recipes from abroad. If you think that a menu like Tomato Vichyssoise, Escalope de Veau and Tarte aux Bananes would be beyond your capabilities, then have a look at the recipes here. You will see that these exotic-sounding dishes can be quickly prepared at home, using your food mixer, so don't imagine that foreign food can only be enjoyed abroad or in expensive restaurants.

Traditional and Foreign Dishes

From Left to Right, Drop Scones
Shortbread and Potato Cakes

Danovang Salad

Using Liquidiser or Blender

Dressing

5 fluid ounce olive oil
2½ fluid ounce tarragon vinegar
1 level teaspoon salt
1 level teaspoon dry mustard

Salad

2-ounce cheese (Gruyère is ideal)
4-ounce cooked chicken (off the bone)
4-ounce cooked ham
½ firm lettuce
Tomatoes

Place all ingredients for the dressing in the goblet and blend for 15 seconds. Dice the cheese, chicken and ham and shred the lettuce. Toss with sufficient dressing to moisten it. Garnish with tomatoes.

Escalopes de Veau

Using Liquidiser or Blender

4-ounce mushrooms
1 onion (roughly chopped)
½ a green pepper
4-ounce butter (melted)
¼ pint sour cream
1 tablespoon dry sherry
Seasoning
4 escalopes of veal (beaten flat and fried in butter)

Chop the mushrooms, onion and green pepper by dropping them through the hole in the lid of the goblet onto the revolving blades. Add this mixture to the melted butter and fry gently for 10 minutes. Add the cream and sherry and bring just to boiling point. Check seasoning.
Pour the sauce over the escalopes then serve with new potatoes and green salad.
Serves 4.

Yorkshire Pudding

Using Liquidiser or Blender

1 egg
4-ounce flour (sieved)
½ pint milk
1 level teaspoon salt
1-ounce lard or dripping

Place all ingredients except the dripping in the goblet and blend for 30 seconds. Leave to stand in a cool place if possible for up to 1 hour.
Heat the fat in a 7 in. × 11 in. tin or individual deep bun tins in a very hot oven (425°F/Reg 7). Cook for 40-50 minutes for the large tin or 15-20 for the small puddings or until brown and crisp. This batter can be made into a sweet pudding by adding fruit to the batter and cooking in lard not dripping.
Serves 6.

Lemon Curd

Using Juice Extractor

4 large lemons
12-ounce castor sugar
½-pound butter
4 eggs

Using a very fine grater take the zest off the lemons onto the sugar. Place these in a large heat resistant bowl. Squeeze out all the lemon juice using the juice extractor, add this to the sugar, lemon rind and butter.
Whisk the eggs lightly, then pour this through a strainer. Add this to the rest of the ingredients. Place the bowl over a saucepan of simmering water and cook, stirring occasionally. Cook until the mixture is thick enough to coat the back of a spoon. Pour into warm jars and cover immediately.
Makes 1½-pounds approximately.

Tomato Vichyssoise

Using Liquidiser or Blender

½-pound onions (peeled and roughly sliced)
¾-pound potatoes (peeled and roughly sliced)
2-ounce butter (melted)
1-pound tomatoes
1 tablespoon tomato purée
1 pint stock
Good pinch of powdered tarragon
Salt and pepper
¼ pint double cream

Place the sliced onions and potatoes in a saucepan with melted butter. Cook gently for 15-20 minutes making sure that they do not brown.
Add the tomatoes, tomato purée, stock and seasonings and simmer gently for about 1 hour or until the vegetables are soft. Allow the mixture to cool slightly. Pour the contents of the saucepan into the goblet and blend until smooth (approximately 30 seconds). Add water or stock at this stage if the consistency is too thick. Chill in the refrigerator. Just prior to serving, stir in the chilled cream. Serve with chervil or chives.

Eierstangen (Egg Sticks)

Using Liquidiser or Blender

2 tablespoons Béchamel Sauce (page 45)
2 hard-boiled eggs (quartered)
Breadcrumbs (made in liquidiser)

Place the sauce plus eggs in the goblet and blend for 10 seconds. Leave the mixture in the refrigerator until it becomes hard enough to roll into small croquette shapes. Roll these in breadcrumbs, place on a lightly greased baking sheet and bake at 400°F/Reg 6 for 20 minutes or until brown.
Makes 6-8 egg sticks.

Liver Pâté

Using Liquidiser or Blender

½-ounce butter (melted in deep saucepan)
½-pound liver (cut into pieces)
2-ounce bacon (cut into pieces)
1 small carrot (roughly chopped)
1 small onion (roughly chopped)
1 stalk of celery (roughly chopped)
1 bay leaf
Salt and pepper
1 tablespoon red wine
½ level teaspoon ground nutmeg

Glaze
¼-ounce gelatine ⎱ heat gently together and
¼ pint meat stock ⎰ then cool slightly

Add the liver, bacon, vegetable pieces and bay leaf to the melted butter, cover the saucepan and cook over a very low heat until the meat and vegetables are tender. Remove the bay leaf and season. Place the contents of the saucepan in the goblet together with the wine and nutmeg—blend until smooth. It may be necessary to stop the motor to push the food back onto the blades. Form the pâté into a loaf shape, coat with gelatine dissolved in the stock. Chill thoroughly before serving.
Serves 6-8.

Frikadeller (Meat Balls)

Using Table Mixer with Mincer and Liquidiser

½-pound raw beef or pork or a mixture of
  the two
1 small onion
1-ounce fine breadcrumbs (made in
  liquidiser)
3 fluid ounce soda water
1 egg
Pinch of salt
Shake of pepper
Butter for frying

Cut the meat into ribbons and then these should be fed through the fine screen of the mincer; follow with the onion.

Place in the bowl with the breadcrumbs, then soda water, egg and seasoning. The mixture should be light and fluffy.

Fry spoonsful of the mixture in the butter until brown on both sides.

These can be served hot with vegetables or cold, sliced, for open sandwiches.

Makes 6 meat balls.

Tarte aux Bananes

Using Table or Hand Mixer

1 (7-inch) sweet pastry flan case (page 77)
1 large banana (sliced into rings)
1-ounce castor sugar
2 tablespoons apricot jam
¼ pint double cream (whipped)

Apricot Glaze

1 tablespoon apricot jam ⎫
1 tablespoon water ⎬ boiled together and sieved

Place the sliced banana in the grill pan, sprinkle with castor sugar and brown under the grill until they just begin to change colour.

Spread the apricot jam over the bottom of the flan case, place half the banana slices on this. Add the whipped cream to fill the flan. Arrange the rest of the banana on top, then brush over the cold apricot glaze.

Serves 6.

Far Left, Frikadeller

Left, Tarte aux Bananes

Sandcake

Using Table Mixer

4-ounce butter
4-ounce castor sugar
2 eggs (separated)
1 tablespoon rum
4-ounce cornflour (sieved)

Beat the butter until soft and pale, add the sugar and then the egg yolks one by one. Continue beating on a medium speed for 5 minutes before beating in the flour and rum. Lastly fold in the stiffly beaten egg whites.

Cook in a deep 7 in. greased cake tin at 350°F/ Reg 4 for 50-60 minutes.

This cake is sometimes preferred slightly under-cooked.

Scottish Oat Cakes

Using Table Mixer

$\frac{1}{2}$ pint water ⎤
3-ounce lard ⎦ together in saucepan

$\frac{3}{4}$-pound oatmeal ⎤
4-ounce self raising flour ⎮ in bowl
1 level teaspoon salt ⎮
2 level teaspoons sugar ⎦

Bring the lard and water to the boil. Add this to the dry ingredients in the bowl and mix on a low speed.

Divide the mixture into 4. Roll each quarter into a round $\frac{1}{3}$ in. thick. Cut each round into 6 segments. Bake at 350°F/Reg 4 on a lightly greased tray for 25-30 minutes or until crisp.

Serve with butter and honey.

Makes 24.

Old-Fashioned Gingerbread

Using Table or Hand Mixer

9-ounce plain flour ⎤
2 level teaspoons ginger ⎬ sieved together
$\frac{1}{2}$ level teaspoon mixed spice ⎦
7-ounce butter or margarine
3-ounce soft brown sugar
6-ounce golden syrup (3 rounded tablespoons)
2-ounce black treacle (2 level tablespoons)
1 egg (beaten)
$\frac{1}{4}$ pint hot water
1 level teaspoon bicarbonate of soda
1 tablespoon cold water

Place the sieved ingredients in the bowl. Melt the butter, brown sugar, golden syrup and black treacle in a saucepan over a low heat. Do not boil. Pour over dry ingredients and mix thoroughly on speed 2.

Add the egg and hot water whilst continuing to beat on the same speed. Dissolve the bicarbonate of soda with 1 tablespoon cold water. Add to the gingerbread mixture and beat for one minute. Pour into a 7 in. square tin previously greased and lined. Bake in a moderate oven 350°F/Reg 4 for $1\frac{1}{4}$-$1\frac{1}{2}$ hours. Allow to cool in the tin for a few minutes, then turn onto a wire rack. This gingerbread is best stored before cutting.

Sour Cream Sauce

Using Liquidiser or Blender

1 raw egg yolk
$\frac{1}{2}$ level teaspoon sugar
$\frac{1}{2}$ level teaspoon salt
Shake of pepper
$\frac{1}{2}$ teaspoon mild mustard
2 tablespoons lemon juice
4 fluid ounce double cream
1 hard-boiled egg (quartered)

Place all ingredients except the hard-boiled egg in the goblet and blend for 10 seconds. Add the egg and blend for a further 2-3 seconds.

Chill before serving with fish or any seafood dish.

Veiled Country Lass

Using Liquidiser or Blender

A delicious apple sweet with a difference

1 (7-ounce) packet of pumpernickel
  (crumbed in liquidiser)
1-ounce sugar
1¼-ounce butter (melted)
8 fluid ounce apple purée made from
  1-1½-pounds stewed apple
Raspberry jam
¼ pint double cream (whipped)

Mix the pumpernickel crumbs with the sugar. Fry this mixture gently in the butter until the crumbs are crisp. Leave to one side.
Spread a layer of apple purée in the base of a serving dish. Top this with a layer of jam and then sprinkle liberally with crumbs. Repeat this pattern until the ingredients are used, finishing with a layer of crumbs.
Top with whipped cream. Chill very thoroughly before serving.
Serves 6.

Kaiserschmarren (Emperors' Pancake)

Using Liquidiser or Blender

½ pint milk
4-ounce plain flour
2 eggs (separated)
1-ounce sugar
1-ounce butter (softened)
1-ounce raisins
Butter for frying

Place the milk, flour, egg yolks, sugar and butter in the goblet and blend for 30 seconds. Add the raisins.
Whisk the egg whites until they hold soft peaks and carefully fold the mixture into them.
Melt the butter in an omelette or small frying pan. Divide the mixture into 2, cooking one half at a time. Pour the batter into the butter, allow to brown on one side then turn over to brown the other.
Sprinkle with castor sugar and pull apart with 2 forks before serving.
This is delicious served with lightly whipped cream.
Serves 2.

Tea Barm Back

Using Table Mixer

½-pound soft brown sugar
1-pound mixed fruit (currants, raisins, sultanas,
  dates, nuts, cherries etc.)
6 fluid ounce black tea
1 egg
10-ounce self raising flour (sieved)

Soak the sugar and fruit in tea overnight.
Place the egg in the bowl and beat until fluffy. Reduce to speed 1 and add the fruit mixture and flour. Bake in a greased 8 in. square tin for 1 hour at 350°F/Reg 4.

Potato Cake

Using Table or Hand Mixer

1-pound cooked potatoes
1-ounce butter
1 tablespoon milk
½ level teaspoon salt
4-ounce plain flour

Place the hot potato, butter, milk and salt in the bowl and beat on speed 3 until the mixture is fluffy. Switch to speed 1, add the flour to form a smooth dough. Shape into flat cakes and fry to serve with egg and bacon or use to make Potato Apple Cake (page 126).

Weinschnitten (Wine Slices)

Using Table or Hand Mixer, or Blender

4 slices bread
1 egg
5 fluid ounce red wine
Sugar and cinnamon

Remove the crusts from the bread and cut each one in half.
Whisk the egg and wine until a smooth liquid is achieved or place the two ingredients in the goblet and blend for 30 seconds.
Soak the bread in the liquid until it is absorbed. Fry in deep fat, then sprinkle with a mixture of sugar and cinnamon. Serve with a wine sauce (Sabayon, page 47) or spread with apricot jam.
Serves 4.

Cacen Berwi (Boiled Cake)

Using Table or Hand Mixer

½ pint milk
4-ounce butter or margarine
6-ounce sugar   } in a saucepan
10-ounce mixed fruit
3 level teaspoons mixed spice
2 level teaspoons bicarbonate of soda
1 egg
12-ounce self raising flour
3 tablespoons milk

Heat the contents of the saucepan to boiling point and allow them to simmer for 1 hour. Remove from heat and stir in the bicarbonate of soda. Leave until cold.
Beat the egg into the cooled mixture on speed 3, then fold in the flour and milk on speed 1.
Transfer the cake mixture to a lined 8 in. square cake tin. Cook at 350°F/Reg 4 for 1½ hours.

Drop Scones

Using Liquidiser or Blender

8-ounce self raising flour
½ level teaspoon salt
2-ounce castor sugar
1 rounded tablespoon golden syrup
1 egg
2-ounce butter or margarine
½ pint milk

For Brushing
½-ounce lard

Place all ingredients in the goblet and blend until perfectly smooth (approximately 1 minute).
Brush a griddle or very heavy based frying pan with a little lard, heat very gently and then spoon small amounts of mixture onto the medium hot surface. When large bubbles have come to the surface, turn and cook the other side until brown. Cool in the folds of a teatowel. Serve with butter.
Makes approximately 16-18.

Danish Pastry

Using Table Mixer

½-ounce yeast
5 fluid ounce warm water } together
1 egg

8-ounce strong plain flour
½ teaspoon salt } sieved together into a bowl
½-ounce sugar
6-ounce unsalted butter

Mix the yeast and warm water together whilst weighing out the other ingredients. Pour this together with the egg into the flour and knead on speed 3 for 3 minutes or until the dough is smooth and leaves the sides of the bowl. Leave covered for 10 minutes.

Soften the butter to the same consistency as the dough.

Shape the dough into a ball and cut a deep cross on the top, (a). Pull each quarter out to form a star shape. Roll these projections out with a rolling pin, (b). Place the fat in the centre and then fold the four corners over the butter. There should now be as much pastry above as below the butter

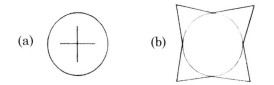

(a)   (b)

Roll the dough out very carefully on a well-floured board into an oblong; mark into three. Fold the bottom third up and the top third down. Give the pastry a half turn and repeat twice, rolling out as thinly as possible and allowing the dough to rest in the refrigerator between each rolling.

After the third rolling, leave the dough to rest in the refrigerator for at least 20 minutes before shaping. After the required shape has been achieved, then leave the pastries for 10-20 minutes in a warm place to prove. Cook at 400°F/Reg 6 for 15-20 minutes until golden brown. Decorate whilst warm.

Note. This pastry can be made one day, stored in the refrigerator overnight, and used the next day.

Suggestions for Shapes

Croissants

Roll the pastry out and cut into 5 in. squares. Cut the squares in half diagonally, then roll from the cut edge to the opposite corner. Place on a lightly greased baking sheet and pull the ends towards each other to form a crescent shape. Glaze with egg and milk. Serve with butter and jam.

Pin Wheels

Roll out the pastry. Cut 4 in. squares, then slash in 1½ in. from each corner towards the centre. Place a teaspoon of filling (see suggestions on page 125) in the centre of each square and pull in alternate pieces from the corners to cover it. When cooked and still warm, brush with Apricot Glaze (page 125), then Glacé Icing. Decorate with nuts.

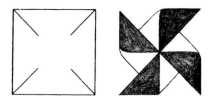

Cocks' Combs

Roll out the pastry and cut into oblongs 4 in. × 5 in. Place a dessertspoon of almond filling on one half, fold the other over this and press well. Slit at ¾ in. intervals. Pull the ends of the comb towards each other so that the slits open. When cooked, brush with apricot glaze and decorate with nuts.

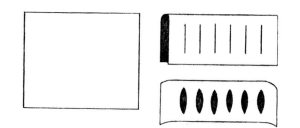

Envelopes

Cut the rolled pastry into 4 in. squares. Put a good teaspoon of vanilla cream (this page) in the centre of each. Bring the corners to the centre and press well. When cooked, finish with Glacé Icing (this page) and nuts.

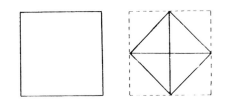

Cushions

Cut the rolled pastry into 4 in. squares, place filling (e.g. half an apricot) in the centre. Bring two opposite corners to the centre.

When cooked, brush with Apricot Glaze (this page) and decorate with nuts.

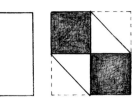

Suggestions for Fillings

Almond

2-ounce ground almonds
2-ounce castor sugar
A little beaten egg

Mix all ingredients together to form a firm paste.

Vanilla Cream

$\frac{1}{4}$ pint milk
$\frac{1}{2}$-ounce cornflour
1-ounce sugar
1 egg yolk
$\frac{1}{2}$ teaspoon vanilla essence

Place all ingredients in the goblet and blend for 10 seconds. Transfer the contents of the goblet to a saucepan, and bring to the boil stirring continuously. Cool before using.

Other Fillings

Jam
Lemon Curd
Fresh or tinned fruit
Fruit purée

Decorations

Apricot Glaze

2 tablespoons apricot jam ⎫ in a saucepan
4 tablespoons water ⎭

Bring the contents of the saucepan to the boil and keep warm.

Glacé Icing

Make up a thick icing, using icing sugar and water.

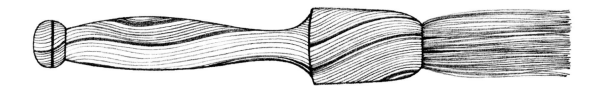

Roscon (*A ring cake with a difference*)

Using Table or Hand Mixer

8-ounce ground almonds
4-ounce castor sugar
6 eggs (separated)
½ level teaspoon cinnamon
Grated rind of ½ lemon

Syrup

2 tablespoons castor sugar ⎤
4 tablespoons water ⎬ boiled for 2 minutes
1 egg white (lightly beaten) ⎦

Mix the ground almonds, sugar, cinnamon and lemon rind, then work in the egg yolks on a medium speed.
Whisk the egg whites until they hold soft peaks, then fold into the almond mixture. Cook in a 1 pint well greased ring mould at 350°F/Reg 4 for 40-50 minutes. Leave to cool for a few minutes before turning it out. Brush with syrup then egg white. Return to the oven for 2-3 minutes.
This can be eaten either hot or cold.

Fröhliche Jungfrauen
(Jolly Spinster Fritters)

Using Table or Hand Mixer

Grated rind of 1 lemon
2-ounce castor sugar
4 eggs (separated)
2 rounded tablespoons flour ⎤
⎬ sieved together
2-ounce cornflour ⎦

Add the lemon rind to the sugar and beat this into the egg yolks on a medium speed until the mixture is pale and creamy in colour. Whisk the egg whites until soft peaks are reached. Fold in the flours on minimum speed, then the whisked egg whites.
Drop spoonsful of the mixture into the hot deep fat and fry until golden brown. Serve with a wine sauce.
Note. This is also ideal as a batter for fruit fritters.
Serves 4.

Shortbread

Using Table Mixer

4-ounce plain flour (sieved)
2-ounce ground rice (or Semolina)
4-ounce butter or margarine (softened)
2-ounce castor sugar

Place all the ingredients in the bowl and mix on minimum speed until they are combined. Increase to speed 3 and mix for a few minutes until the shortbread becomes soft and pliable. Roll out on a lightly floured board until approximately ¼ in. thick. Prick well and cut out into the desired shapes.
Place on a lightly greased sheet and bake at 325°F/Reg 3 for 25-30 minutes, or until pale gold in colour. Sprinkle with castor sugar and store in an airtight tin when cold.
Alternately press the mixture into a lightly greased 8 or 8½ in. tin, prick and cook at the same temperature for 30-45 minutes. Cut the round into 12 pieces and cool before storing in an airtight tin.

Potato Apple Cake

Using Table or Hand Mixer

1 recipe of Potato Cake (page 122)
½-pound cooking apples (peeled and sliced)
2-ounce butter
2-ounce castor sugar

Roll out the Potato Cake dough on a lightly floured surface to ½ in. thickness. Cut into 4 squares. Cover half of each square with apple, dampen the edges and then turn the other half over to form a triangular turnover. Place on an ungreased baking sheet and bake at 400°F/Reg 6 for 30 minutes or until golden brown. Lift up the top and place in the butter and sugar. Sprinkle the top of each with castor sugar.
Serve either hot or cold.

INDEX